Proceedings of the First International Symposium on Deep-Sea Corals

CONSENSUS STATEMENT ON THE NEED TO CONSERVE DEEP SEA CORALS

The First International Symposium on Deepsea Corals, covering the biology, ecology, geology, conservation and management of deepwater corals and their environment, was held in Halifax, Canada, 30 July - 3 August, 2000. The following is a consensus reached at a conservation workshop held in association with the Symposium, which was attended by scientists from Europe, North America and the Southern Hemisphere, as well as by fishermen and representatives of NGOs, government and international agencies.

Corals are now recognized to be as diverse in the deep seas and oceans as in shallow tropical waters. They form reefs and other benthic structures, often in association with complex topography, along continental margins, on seamounts and ridges, through much of the world's oceans.

Deepwater corals typically exhibit slow growth and extreme longevity. The coral species and their associated communities may have highly localized distributions. As a consequence of these features, they are exceptionally vulnerable to disturbance.

Seamounts and other coldwater coral habitats are today subject to growing anthropogenic pressures, in particular as a result of the rapid spread of deepwater trawl fisheries into new regions and new grounds, aided by the explosive development of mapping, navigational, fish-finding, and other technologies.

The role of deepwater coral species and reefs in forming essential habitats for fishes and other species, and their potential for biotechnology and pharmaceuticals, are still poorly understood, but their exceptional diversity, uniqueness and vulnerability necessitates that their conservation be given high priority.

Steps to conserve and restore deepsea coral and reef environments need to be taken at the national, regional and international levels to protect deepsea coral reefs and areas of high biodiversity associated with corals, both within areas of national jurisdiction and on the high seas. The establishment of deepwater reserves should follow the best principles of ecology and of sustainable, precautionary management, such that a global and representative system of marine protected areas is established which is adequate for the long-term conservation of deepwater biodiversity. It is essential that existing national laws and international conventions for the protection of biodiversity and the regulation of fisheries be extended to cover these unique and vulnerable deepwater habitats.

Proceedings of the First International Symposium on Deep-Sea Corals

Edited by

J.H. Martin Willison, June Hall, Susan E. Gass,
Ellen L.R. Kenchington, Mark Butler and Penny Doherty

This volume contains a selection of the papers presented at a conference organized by the Ecology Action Centre and held at Dalhousie University in Halifax, Nova Scotia, July 30 to August 3, 2000.

Ecology Action Centre
Nova Scotia Museum
Halifax, Nova Scotia
2001

With the generous assistance of:
The Henry P. Kendall Foundation
The Pew Charitable Trusts
The Marine Conservation Biology Institute
Fisheries and Oceans Canada

Front cover illustration of *Primnoa* coral and habitat by Susan Feindel. Courtesy Mark Butler
Back cover photos: Bedford Institute of Oceanography, Department of Fisheries and Oceans and Derek Davis

National Library of Canada Cataloguing in Publication Data

International Symposium on Deep-Sea Corals (1st: 2000: Dalhousie University)
Proceedings of the First International Symposium on Deep-Sea Corals

(Marine Issues Committee Special Publication; 10)
Conference held at Dalhousie University, Halifax, N.S., July 30-August 3, 2000.
ISBN 0-9683068-7-X

1. Corals—Congresses. 2. Deep-Sea ecology—Congresses. 3. Coral reef ecology—Congresses. 4. Coral reef conservation—Congresses. 5. Deep-sea animals—Congresses. I. Willison, J.H. Martin II. Ecology Action Centre. III. Title. IV. Series.

QL 377.C5I58 2000 593.6 C2001-904135-7

To obtain copies of this report, contact one of the following:

Ecology Action Centre
1568 Argyle St., Suite 31
Halifax, Nova Scotia
Canada B3J 2B3
phone: (902) 429-2202
fax: (902) 422-6410
email: eac_hfx@istar.ca

Nova Scotia Museum of Natural History
1747 Summer St.
Halifax, Nova Scotia
Canada B3H 3A6
fax: (902) 424-0560
email: nsmbooks@gov.ns.ca

Printed in Canada

Contents

DEDICATION

This volume is dedicated to the memory of Dr. Donald E. McAllister, who died on June 17th, 2001. Don attended the First International Symposium on Deep-Sea Corals and is an author of a paper in this volume. He was a gentle, soft-spoken man with a huge heart both for humanity and the creatures of the sea. Don worked at the Canadian Museum of Nature in Ottawa. He was chief editor of the journal *Global Biodiversity*, published initially by the museum, and then continued under the title *Biodiversity*, published by the organization Tropical Conservancy (www.synapse.net/~tropical). Don is best known for his tireless volunteer work. He established Ocean Voice International and worked constantly to promote its goals of fairness and conservation. The journal *Sea Wind*, published by Ocean Voice since 1987, and edited by Don, carried the banner "Toward harmony of humankind, the sea and its life." Don carried this message in his heart at all times; he stood equally for social and environmental justice. The world is a poorer place without him. For information on Ocean Voice International, go to www.ovi.ca/.

Ellie Dorsey, who helped to organize publicity for the symposium, died shortly before it was held. The symposium was dedicated to her memory with the words "Ellie Dorsey was a cheerful and intelligent advocate for ocean conservation." Ellie was working with COMPASS (Community Partnership for Science and the Sea) when she died. Ellie and COMPASS successfully helped to ensure that the deep-sea coral conservation messages heard at the symposium were communicated to a wider world through the popular media.

INTRODUCTION

Julia Whitty wrote about threats to the world's coral reefs in the January 2001 issue of *Harper's Magazine*, published in New York. In her article, "Shoals of Time," she stated: "Although many of the 6,000 species of corals and anemones are solitary polyps, or colonies of soft corals, or those that build flat calcium-carbonate formations in the seemingly inhospitable waters of the Arctic and Antarctic, the story of coral reefs occurs in the narrow band of equatorial water at the 21°C isotherm."

By contrast with the above, in October 1999 under the headline "Britain's secret coral reefs at risk," reporter Anthony Browne wrote as follows in *The Observer*, a British newspaper: "Scientists have confirmed that the British Isles are surrounded by hundreds of deep-sea coral reefs up to 400ft high and possibly miles long, with as many species living in them as a tropical rainforest." As anyone who has put a toe in the seas around Britain will quickly confirm, these seas are definitely not "equatorial water."

The Proceedings of the First International Symposium on Deep-Sea Corals was put together in the context of the current popular interest in, and popular misunderstanding of, life in the deep seas. Through this volume we hope to let a wide audience know that coral reefs are found in cold ocean waters, that corals are abundant in both Canada and the United States, and that threats to them are very real.

HOW WE GOT STARTED

The earliest known records of deep-sea corals in the Atlantic region of Canada came from fishermen and scientific expeditions in the 1800s. Remarkably, local interest in these organisms then declined, and some knowledge may have been lost. In the late 1990s there was a revival of interest in the subject locally when people in several walks of life began to try to understand the causes of collapses in regional fisheries, which had reached crisis proportions in the early 1990s.

In 1996 a group of fishermen in southern Nova Scotia created the Canadian Ocean Habitat Protection Society (COHPS). One of their main messages was that destruction of coral "trees" was among the reasons for failing fisheries. Generations of Nova Scotia fishermen had understood that these strange "trees" were important to their livelihoods. As members of the Ecology Action Centre, an environmental organization in Halifax, Nova Scotia, we decided to assist the fishermen of COHPS in their work. The fishermen were pleased to have this help because they had been unable to persuade fishery managers that this was an important matter, and they thought a broader coalition would help their cause.

In 1997 the Ecology Action Centre received a grant from the Nova Scotia Museum of Natural History to conduct a study on the distribution and status of deep-sea corals around Nova Scotia. The work was conducted by Heather

Breeze and was published later that year by the Marine Issues Committee of the Ecology Action Centre. Most of the information was gathered from fishermen, but in addition all museum and scientific records were collated and mapped. In association with this research, we contacted scientists in several countries where deep-sea coral research had been conducted, and similar stories kept recurring. We found that fishermen in many parts of the world were reporting that regions where corals had once been found had been cleared of them. Reports of scientific studies on deep-sea corals were also received from Norway, Australia, Alaska and Germany, several of which reported damage done to coral communities by fishing.

We realized that a lot of research on deep-sea corals was going on in many countries outside Canada, and there was a lot of interest in the subject scattered around the world. Nevertheless, we were surprised to discover that no worldwide scientific conference had been held on the subject. Except for the comprehensive book on Norway's coral reefs by Martin Hovland and Pål Mortensen, published in 1999 by John Grieg Forlag (Bergen), we couldn't even find a book dedicated to the subject. As a result, perhaps naively, we decided to take on the tasks of holding a meeting and publishing a book.

When we planned the First International Symposium on Deep-Sea Corals, we hoped to bring experts from around the world to Nova Scotia. This was intended partly to help our own local attempts to raise awareness about the importance of the local corals. On a larger scale, we thought that the meeting might have beneficial effects all around the world, because our situation was clearly not unique. At the outset, we also realized that multi-disciplinary conventional science standing alone would be insufficient to affect marine management policies, and so we decided to bring people together from a range of social sectors: scientists, artists, activists, administrators, artisans and business people. Ultimately we were delighted with the success of the whole venture: lots of valuable information was exchanged and in Canada, at least, some progress has since been made towards better conservation of deep-sea corals. We hope that this volume will contribute further to this progress, and over a wider geographic area.

THE PROCEEDINGS OF THE FIRST INTERNATIONAL SYMPOSIUM ON DEEP-SEA CORALS

In this volume, we have tried to reflect the multi-dimensional nature of the symposium, and also produce a work which will inform those who want to help re-educate the public about corals. In order to be fully informed, the reader is recommended to find the special issue of the scientific journal *Hydrobiologia*, to be published in 2002, which contains further papers from the symposium that are included here only as abstracts (see pp. 194-204).

The symposium papers in this volume have been organized to reflect three themes: deep-sea corals themselves; human threats to deep-sea corals; and conservation of deep-sea corals, including popular appreciation of them. The story told as these themes unfold is as follows.

Corals, whether in shallow tropical marine waters or in the deep seas, face many threats, notably: changes in temperature, burial by sediments, pollution, and physical disruption by fishing gears. For deep-sea corals, however, there is the additional insidious threat of public ignorance. It's difficult to care about something unknown. The first major steps to overcoming this ignorance came from the work of scientists in Europe, who in recent years have revealed to one surprised European nation after another the existence of "secret coral reefs."

The stony coral *Lophelia pertusa* was originally named by Linnaeus in 1758, but it was only in the 20th century that scientists in Europe began to realize it forms large mounds which can properly be described as "reefs." Previously, this coral had been assumed to be another of the myriad little invertebrate creatures that attach themselves here and there to suitable substrate on the bottom of the seas and oceans. Many hundreds of *Lophelia* mounds have now been found in several regions of the continental shelves of Europe, from the territories of Portugal, Spain, Ireland, Great Britain, Sweden and Norway, as far north as the Arctic Circle. In the first paper in this volume, John Wilson provides us with an original view of the revolutionary discovery of Europe's coral reefs.

The accumulated mound of calcium carbonate made by reef-forming corals is sometimes called a "bioherm," a word having Greek roots and meaning "life-mound," an apt name for a coral reef. In the volume's second paper, Bruce Hatcher and Robert Scheibling compare the ecological attributes of tropical coral reefs found in relatively shallow waters with the attributes of reefs found in deep ocean waters, including the *Lophelia* mounds of the boreal Atlantic. They conclude that boreal bioherms are indeed "coral reefs," and propose that rigorous studies should be conducted in order to discover whether the well-developed body of ecological theory concerning tropical coral reefs applies equally well to deep reefs in cold water.

Many corals in both shallow tropical waters and the cold waters of deep oceans are not reef-forming. In the third paper, Maya Borel Best provides us with a taxonomist's view of the distinction between those corals that form reefs and those that do not, as well as between those that live in shallow water, and those that do not. While we might assume that these very different locations and life strategies would be adopted by distinctly different animals, she shows that this is not so, using the stony coral genus *Madracis* as an example. Some *Madracis* species are described as "facultatively zooxanthellate," meaning that they can associate with photosynthetic microorganisms and obtain nutrition this way in shallow water, or they can dispense with their zoo xanthellae and live at greater depth in the dark by filter-feeding.

Many hundreds of species of stony corals (Scleractinia) are known to exist, of which a little over half do not form reefs. Most of the non-reef-forming scleractinian corals of the deep seas are described as "solitary," because the individual polyps do not form a collaborative community as they grow in number. Being ancient organisms in an evolutionary sense, corals have had plenty of time to develop the capacity to exploit most kinds of marine benthic environments, limited only by the fact that the stony core of the adults makes them heavy and therefore stuck on the bottom, though like most marine invertebrates, their eggs float freely and are dispersed in ocean currents. In the fourth paper, Stephen Cairns and Ralph Chapman tell us about the distributions of 134 of these species in the cold, deep waters of the North Atlantic. They show that there are three major spatial "clusters" based on biogeographic affinities. In every major region there are "endemic" species, i.e., species that live there and are known from nowhere else.

Specimens of deep-sea corals are obtained by many means, often coming up on fishing gears, and sometimes in sampling devices deployed by scientists. Observing them *in situ* is of special interest, for without such observations our knowledge of their ecology is very limited. In the next paper, Kevin MacIsaac and colleagues from Canada's Bedford Institute of Oceanography in Nova Scotia describe their searches for corals on the Scotian Shelf, to the east of Nova Scotia. They attempt to define the habitat preferences of the four major groups of corals found there: stony (Scleractinia), horny (Gorgonacea), soft (Alcyonacea) and pen (Pennatulacea). Leonardo Tunesi, Diviacco Giovanni and Giulia Mo similarly describe how they used a submersible vehicle to hunt for scleractinian corals in the Gulf of Tigulio, in the Mediterranean Sea near Italy. Finding only small remnants, many of them dead, the authors hypothesize that past climate change or some other factor has made the area less suitable for those species.

Reef-forming stony corals have received a lot of attention, but they are not the only corals that contribute significantly to the three-dimensional structure of the bottom of the ocean. Gorgonacean corals can form large tree-like colonies attached to rocks or shells, and these have been observed to provide shelter for fish and other marine species in a similar manner to the creation of habitat by tropical corals for a myriad of other life forms. Little is known about the manner in which the gorgonacean corals recruit to new sites, or about the rates at which they grow in deep cold waters. In the next paper in the volume, Robert Stone and Bruce Wing describe the methods they used to assess the growth rate of the Alaskan horny coral, *Calcigorgia spiculifera*.

In order that our understanding of deep sea corals increases and is put to good use, large integrated research programs will be needed in several regions of the world. The Atlantic Coral Ecosystem Study (ACES), described by Anthony Grehan and André Freiwald, focuses on the eastern part of the Atlantic Ocean and is being conducted using funds from the European Union. It is a potential model for other such studies. ACES aims to assess the vulnerability of Europe's coral reefs, and provide conservation advice for those making decisions about industrial development in the region.

It is well recognized that the use of mobile fishing gears is a major threat to deep-sea corals. Large protruding organisms that are fixed to the bottom are clearly specially vulnerable to the huge heavy trawls widely used in bottom fishing. Around the margins of the continental shelves of the North Atlantic, trawling has reached depths of 2000 metres as fish stocks have been depleted and fishing companies have sought new stocks of fish to exploit, yet outside of Europe few studies have been conducted on the impacts of these fishing techniques on deep-sea corals. A significant exception to this are the observations of Kenneth Krieger, who describes a remarkable study conducted in Alaska. In 1990, a National Marine Fisheries Service carried out a routine research trawl survey through two regions in which the gorgonacean coral *Primnoa resedaeformis* was plentiful. Mobile-gear fishing had never been conducted at one of these sites, which was essentially pristine. A ton of coral came up in the trawl after a single trawl sweep lasting 30 minutes. Seven years later, the sites were examined using a submersible vehicle, and the extent of the damage caused by the research trawling was assessed. The trawl paths were still clearly visible, and at one of the sites there was no evidence that young corals had recruited to the damaged site.

In view of the damage caused to Alaskan deep-sea corals by fishing gears, specific conservation measures are being examined by the North Pacific Fishery Management Council. In their paper, David Witherell and Cathy Coon discuss options for the protection of thickets of Alaskan corals. Using data from fishery trawl surveys, the distribution of large corals was mapped and marine protected areas were proposed, but fishermen questioned the accuracy of the mapping and the proposals were tabled. Nevertheless, there was general agreement that corals are important fish habitat and should be protected, and also that improved mapping would make this possible.

The most common response of conservationists to a realization that specific types of natural communities are at risk is to propose that protected areas be established. Those activities which threaten to radically alter the natural community are banned in protected areas. The approach is regarded by many as an insurance policy, established to ensure that the community type is retained if other conservation measures fail outside the protected areas. Since deep-sea corals constitute a foundation for complex benthic communities that are specially vulnerable due to the fragile nature of this foundation, it is very important to know as much as possible about where the corals are located. Even fragmentary information is of value. Don McAllister and Noel Alfonso report their examination of museum records and data from fisheries observers for the British Columbia region of the Pacific. Gorgonacean, alcyonacean, scleractinian and antipatharian corals have been reported at depths between 20 and 2000 metres in waters adjacent to British Columbia.

In theory, protected areas are best planned as systems, based on the intention to include representative examples of all of the diversity of habitat types (biotopes) found within any region for which planning is being undertaken. Christopher Miller examines the Scotian Shelf as a case study, paying particular attention to the attributes of deep-sea corals that make them of special interest for conservation in a marine protected area strategy. Marine protected areas will be a suitable means to protect the best aggregations of corals, but this strategy alone will not be sufficient to protect all deep-sea corals because some thickets are bound to be found in areas which are fished. Mark Butler and Susan Gass examine coral conservation strategies for fished regions, such as rating fishing gears according to their destructiveness and developing disincentives to discourage the use of relatively destructive gears in sensitive habitats, such as where corals are present. They also discuss the regulation of other industrial activities, such as petroleum mining, in relation to marine habitat sensitivity.

One of the regions which Miller identifies as specially important is the Northeast Channel lying between Browns Bank and Georges Bank on the Scotian Shelf. This area has been known as a coral hotspot by generations of Nova Scotia fishermen. Recent interest in the subject of deep-sea corals in Nova Scotia arose as a result of the work of a group of fishermen in southern Nova Scotia, some of whom fished the Northeast Channel region. Their concern for the protection of "the trees," as they call the gorgonacean corals found there, led to the creation of the Canadian Ocean Habitat Protection Society (COHPS). Derek Jones, who is one of these fishermen, and Martin Willison describe the work of COHPS, and how their work led eventually to the symposium on which this volume is based.

Education is a major part of the work of COHPS. Their interest in promoting better understanding of local corals led to the development of a "Corals and Canyons" education kit by the Ecology Action Centre and the Nova Scotia Museum, both located in Halifax. In the penultimate chapter, Penny Doherty and others describe the kit and its use.

Many people find corals to be very beautiful, and COHPS has used their visual appeal as an advocacy tool. Susan Feindel, an independent graphic artist living in Nova Scotia, was captured on hearing the story of the destruction of Nova Scotia's coral forests. A lot of her previous work had focused on understanding land, as a geologist might, and human impacts on it, as an environmental scientist might. This interest led Susan to join a marine scientific cruise and to observe the work of scientists as they sought to collect information about corals on the Scotian Shelf. In the final paper, she describes her work and provides us with unique graphic glimpses of art in science, around the theme of corals and underwater landscape.

Immediately following the symposium, a one-day workshop was held so that scientists and conservationists could get together and plan actions that might be taken to promote a global conservation agenda for deep-sea corals. June Hall's report on the workshop is provided at the end of the volume, as are abstracts of all other oral and poster presentations, and a full list of symposium participants and their addresses.

Martin Willison and Susan Gass
Dalhousie University and Ecology Action Centre
Halifax, Nova Scotia
August 2001

ACKNOWLEDGMENTS

The editors would like to thank the Marine Conservation Biology Institute, the Pew Charitable Trusts, and the Henry P. Kendall Foundation for their support in the development and organization of the symposium and the publication of the proceedings. We would also like to thank the Department of Fisheries and Oceans for their support and for their assistance with the distribution of the proceedings.

Other organizations that provided support for the symposium are the Canadian Ocean Habitat Protection Society, the Community Partnership for Science and the Sea (COMPASS), Dalhousie University Instructional and Media Services and Conference Services, DFO Oceans and Science Branch and Science and Technology Youth Internship Program, the Federation of Nova Scotia Naturalists, the Science and Management of Protected Areas Association, and the World Wildlife Fund (Canada).

Special thanks are due to Susanna Fuller, Andrew Hebda, Inka Milewski and Christine Patterson for their help in planning the symposium; to Symposium Coordinator Susan Gass for help with the administrative aspects of publication; and to copy-editor June Hall.

Finally, we would like to thank the members of the Symposium Science Committee for their invaluable contributions to the symposium planning process: Stephen Cairns, Paul Dayton, Jan Helge Fosså, André Freiwald, Donald Gordon, Bruce Hatcher, Tony Koslow, Don McAllister, Michael Risk, Murray Roberts, Dave Scott and Les Watling.

LOPHELIA 1700 TO 2000 AND BEYOND

John B. Wilson
Department of Geography, Royal Holloway University of London, Egham, Surrey, UK

SUMMARY

The earliest records of the deep-water coral *Lophelia pertusa* are Norwegian. In 1755 it is referred to as being collected by fishermen and divers and used for medicinal purposes. The species was described in 1758 and 1768. The growth of fishing in the later 18th to early 19th century provided many more records on its distribution from Norway to the Iberian Peninsula. In the 1960s interest widened to New Zealand, where the colonization strategy of *Lophelia* was investigated. Submersible observations were made in the late 1960s and 1970s on the Blake Plateau, in the Straits of Florida and on Rockall Bank. These provided further models for its colonization strategies. Investigations off northern and mid Norway (Stjernsund and Sula Ridge) have provided much new information on the ecology and general geobiology of *Lophelia*.

In the early 1990s interest developed in the potential oil prospects northwest of Shetland in the so-called Atlantic Frontier Zone. This generated wide interest in *Lophelia*, partly in response to inaccurate scare stories about the alleged destructive effects of oil developments on the coral. Many more research institutes and universities became interested in the general ecology, physiology and genetic variability of *Lophelia*, culminating in the start of the European Union ACES (Atlantic Coral Ecosystem Study) project. This research will concentrate on specific topics including an appraisal of the potential destructive effects of extensive continental margin trawling.

LOPHELIA IN THE 18TH AND 19TH CENTURIES

The earliest records of deep-water corals are from Norwegian waters. These early descriptions allude to the corals as 'sea trees' and the associated fauna as 'sea vegetables.' In his account of the natural history of Norway, the Right Reverend Erich Pontoppidan, Bishop of Bergen and Member of the Royal Academy of Sciences at Copenhagen, summarizes the state of knowledge of deep-water corals (certainly including *Lophelia*) in Norwegian waters in the mid-18th century (Pontoppidan, 1755). The corals were known and collected by fishermen and divers and were used partly as food but "likewise for powerful medicaments." These medicinal uses were based on the coral being "absorbent, refrigirative, emollient, astringent and strengthening." The medicines were taken internally as extracted fats and oil. Coral beads worn round the neck were "a preservative against the apoplexy, the plague and other contagions." Coral was traded in Genoa and Marseilles and used to make necklaces and bracelets for the Eastern market. Pontoppidan suggested that if white coral (*Lophelia*) could be brought into fashion then "a diligent search might procure as great a quantity in our seas." Pontoppidan also includes descriptions of collected specimens, some of which are figured, and he comments on the processes of calcification.

Linnaeus formally described the species in the 10th edition of his *Systema Naturae* (Linnaeus, 1758) as *Madrepora pertusa*. Further details of the Norwegian occurrences of *Lophelia* were published in 1768 by Johan Ernst Gunnerus, a later Bishop of Bergen, who describes several specimens in detail and provides figures of some of them. Two figures (his Figures 1 and 2) are the first scientifically accurate drawings of *Lophelia* (published as *Madrepora pertusa* in accordance with the Linnaean classification of the time). Gunnerus refers to active discussions and correspondence with other contemporary workers and to the forthcoming publication of the 11th edition of *Systema Naturae*.

The growth of fishing in the late 18th and 19th centuries provided more records on the distribution of *Lophelia* from Norway to the Iberian Peninsula. In Scottish waters, four specimens were obtained by fishermen in the early to mid-19th century. They are referred to in the classic publications of Johnston (1847) and Gosse (1860), and were located again and figured by Wilson (1979a). It is perhaps significant that these specimens were collected from nearer-shore waters where *Lophelia* is not commonly found today. *Lophelia* was dredged in 1868 from the HMS *Lightning* from a depth of 969 m and from HMS *Porcupine* at a depth of 664 m west of Scotland (Duncan, 1873; Thomson, 1874). The first records from Rockall Bank were from a depth of 300 m in 1869 (Duncan, 1873).

LOPHELIA IN THE 20TH CENTURY

The growth of fishing activity along the northwest European continental shelf and upper continental slope provided a wealth of data on the distribution of *Lophelia* and other deep-water corals (Joubin, 1922a,b; Le Danois, 1948; Dons, 1944; Wilson, 1979b). The results suggest that in the first half of the 20th century *Lophelia* was known to be quite abundant. Joubin (1922a) reports that a fishing vessel on one occasion brought up some 5-6 tons of coral in one of its trawls. The sections through the outer continental shelf and upper continental slope published by Le Danois (1948) show substantial coral banks on the continental slope in the Southwest Approaches, west of Porcupine Bank and off northern Spain.

The accumulated data resulted in several publications including Teichert (1958) and Fernandez Garcia et al. (1976). A very brief summary of other occurrences of *Lophelia* on the west side of the Atlantic and other locations is given by Jurgensen (1916). Some of these papers contain ecological observations. The present paper must necessarily be regarded as a very brief review of some of the papers published in the 20th century.

In the 1960s interest widened to New Zealand, where the colonization strategy of *Lophelia* was investigated (Squires, 1965). Manned submersible observations were made in the last 1960s and 1970s on the Blake Plateau

(Milliman et al., 1967), in the Straights of Florida (Neumann et al., 1977), on Rockall Bank (Wilson, 1979c), Porcupine Bank and the Porcupine Seabight (Tudhope and Scoffin, 1995). These provided further models for the colonization strategies of the species and much ecological and geobiological knowledge was obtained during these investigations.

Recent research off northern and mid Norway (Stjernsund and Sula Ridge), also using manned submersibles and ROVs, has provided much new information on the ecology and general geobiology of *Lophelia* (Freiwald, 1997; Freiwald, 1998, Freiwald et al., 1999; Mortensen et al., 1995) and has established that *Lophelia* can build true reefs when conditions are appropriate. The current state of knowledge regarding the general biology and distribution of *Lophelia* can be found in the review by Rogers (1999) and on the Atlantic Margin in particular by Roberts et al. (in prep.).

In the early 1990s, interest developed in the potential oil prospects northwest of Shetland in the so-called Atlantic Frontier Zone, in the Rockall Trough and on the western slopes of Porcupine Bank. As a result, several detailed and regional surveys of the area have been undertaken; notable findings include the discovery of the so-called Darwin Mounds (Bett, 1999, 2000) in the vicinity of some of the occurrences reported by Wyville Thomson (1874). Many of these mounds support colonies of *Lophelia*. Observations suggest that the colonies are of the order of 30 cm in height. This is comparable to the size of colonies observed on the slopes of Porcupine Bank (Wilson and Vina Herbon, 1998) and on Galicia Bank. It is possible that the relatively small size of these colonies may be related to one of the effects of continuing fishing activities in these areas. They are substantially smaller than the 1- to 2-metre-high colonies observed in 1973 on Rockall Bank (Wilson, 1979c).

The economic potential of the Atlantic Frontier Zone has generated wide interest in *Lophelia*, partly in response to inaccurate scare stories about the alleged destructive effects of oil developments on the coral. Many more European research institutes and universities became interested in the general ecology, physiology and genetic variability in *Lophelia* and this interest culminated in the start of the European Union ACES (Atlantic Coral Ecosystem Study) project in April 2000. Ten European organizations are partners in the project, which is coordinated by Professor André Freiwald of the University of Tübingen. The research commissioned by ACES is organized into five work packages and will concentrate on specific topics including reproductive strategies; growth rates; framework-constructing potential; the longevity of the deep-water coral ecosystem; biodiversity; coral behaviour and sensitivity to stress; hydrographic and other factors affecting benthic boundary layer particle dynamics; and conservation issues including an appraisal of the destructive effects of extensive continental margin trawling.

While these investigations will undoubtedly resolve many of the present unknowns in *Lophelia* research, it is certain that other problems will come to light that will lead to further research in the future.

REFERENCES

Bett, B.J. 1999. RRS Charles Darwin cruise 112C Leg 2, 19 May-24 June 1998. Atlantic Margin Environmental Survey: seabed survey of deep-water areas (17th round Tranches) to the north and west of Scotland. Southampton Oceanography Centre, Cruise Report no. 25, 171 pp.

Bett, B.J. 2000. RRS Charles Darwin cruise 119C. Initial assessment of seabed observations made during the White Zone (WhiZ) Environmental Survey: Seabed survey of deep-waters to the north and west of Shetland. Southampton Oceanography Centre, Preliminary Cruise Report.

Dons, C. 1944. Norges korallrev, Det Kongelige Norske vidskabernes selskabs, Forhandlinger 16, 37*-82*.

Duncan, P.M. 1873. A description of the Madreporaria dredged up during the expeditions of HMS 'Porcupine' in 1869 and 1870, Transactions of the Zoological Society of London 8: 303-344.

Frenandez Garcia, A., S. Iglesias Martinex, F.J. Pereiro Munoz and M. Caloca. 1976. Primer estudio de al pesqueria demersal de Grand Sol y oeste de Irlanda para la flota espanola, Boletin del Instituto espanol de oceanografia 213, 37 pp.

Freiwald, A. 1998. Geobiology of *Lophelia pertusa* (Scleractinia) reefs in the north Atlantic. Habilitation thesis, Fachbereich Geowissenschaften, Universität Bremen, Bremen, 116 pp.

Freiwald, A., R. Henrich and J. Pätzold. 1997. Anatomy of a deep-water coral reef mound from Stjernsund, West-Finnmark, northern Norway, in cool-water carbonates. In N.P. James and J.A.D. Clark (eds), SEPM Special Publication 56: 141-161.

Freiwald, A., J.B. Wilson and R. Henrich. 1999. Grounding Pleistocene icebergs shape recent deep-water coral reefs, Sedimentary Geology 125: 1-8.

Gosse, P.H. 1860. Actinologica Britannica. A History of the British Sea-Anemones and Corals, London.

Gunnerus, J.E. 1768. Om nogle Norske coraller. Det Kongelige Norske vidskabernes selskabs skrifter 4: 38-73.

Johnston, G. 1847. A History of the British Zoophytes. 2nd edition, 2 volumes, London.

Joubin, L. 1922a. Les coraux de mer profonde nuisibles aux chalutiers. Notes et Mémoires. Office scientifique et technique des pêches maritimes 18, 16 pp.

Joubin, L. 1922b. Distribution géographique de quelques coraux abyssaux dans les mers occidentales européenes, Compte rendu hebdomadiare des séances de l'Académie des sciences 175: 930-933.

Jurgensen, H.F.E. 1916. Alcyonarian and Madreporian corals in the Museum of Bergen collected by the "Fram" Expedition 1898-1902 and by the "Michael Sars" 1900-1906. Bergens Museum Aabok Naturvidenskabelig Raekke 1915-1916 No. 6: 1-44.

Le Danois, E., 1948. Les profondeurs de la mer. Payot, Paris, 303 pp.

Linnaeus, C. 1758. Systema naturae per Regna tria naturae, secundum classes, ordines, genera, species. Tomus 1: Regnum animale, 10th edition, Stockholm, 824 pp.

Milliman, J.D., F.T. Manheim, R.M. Pratt and E.F.K. Zarudski. 1967. ALVIN dives on the continental margin off the southeastern United States. Technical Report. Woods Hole Oceanographic Institution 67-80, 48 pp.

Mortensen, P.B., M. Hovland, T. Brattegard and R. Farestveit. 1995. Deep water bioherms of the scleractinian coral *Lophelia pertusa* (L) at 64°N on the Norwegian shelf: Structure and associated megafauna, Sarsia 83: 145-158.

Neumann, A.C., J.W. Koefoed and O. Keller. 1977. Lithoherms in the Straits of Florida, Geology 5, 4-10.

Pontoppidan, E. 1755. The Natural History of Norway. A. Linde, London.

Roberts, J.M., D. Long, J.B. Wilson, P.B. Mortensen and J.D. Gage. The distribution of Lophelia pertusa (Linnaeus, 1758) and other deep-water reef-forming corals and impacts from activities, Internationale Revue de Hydrobiologie 84: 315-406.

Squires, D.F. 1965. Fossil coral thickets in Wairarapa, New Zealand, Journal of Palaeontology 38: 904-915.

Teichert, C. 1958. Cold and deep-water coral banks, Bulletin of the American Association of Petroleum Geologists 42: 1064-1082.

Thomson, C. Wyville.1874. The Depths of the Sea. London, 527 pp.

Tudhope, A.W. and T.P. Scoffin. 1995. Processes of sedimentation in Gollum Channel, Porcupine Seabight: submersible observations and sediment analyses, Transactions of the Royal Society of Edinburgh, Earth Sciences 86: 49-55.

Wilson, J.B. 1979a. The first recorded specimens of the deep-water coral *Lophelia pertusa* (Linnaeus, 1758) from British waters, Bulletin of the British Museum Natural History (Zoology) 36: 209-215.

Wilson, J.B. 1979b. The distribution of the coral *Lophelia pertusa* (L.), [*L. prolifera* (Pallas)] in the north-east Atlantic, Journal of the Marine Biological Association of the United Kingdom 59: 149-164.

Wilson, J.B. 1979c. 'Patch' development of the deep-water coral *Lophelia pertusa* (L.) on Rockall Bank, Journal of the Marine Biological Association of the United Kingdom 59: 165-177.

Wilson, J. and C. Vina Herbon. 1998. Macro faunas and biogenic carbonates from the north slope of Porcupine Bank, south east slope of Rockall Bank and the west of Faeroe Bank Channel. In N.H. Kenyon, M.K. Ivanov and A.M. Akhmetzhanov (eds), Cold water carbonates and sediment transport on the Northeast Atlantic Margin, UNESCO Intergovernmental Oceanographic Commission Technical Series 52: 149-157.

WHAT DETERMINES WHETHER DEEP-WATER CORALS BUILD REEFS: DO SHALLOW REEF MODELS APPLY?

Bruce G. Hatcher
Robert E. Scheibling
Biology Department, Dalhousie University, Halifax, Nova Scotia, Canada

SUMMARY
Geologists view coral reefs as carbonate accumulations produced by corals ("bioherms"). Little matter whether the surface veneer of living corals is present: the coral assemblage will come and go with environmental change, but the reefs can persist for millennia. Ecologists see coral reefs as marine communities structured by living corals ("coral reef ecosystems"). It matters little whether the corals attach to dead coral rock: corals will recruit to any hard substratum, but reefs can only develop when coral communities persist. Three attributes of coral assemblages permit the formation of benthic communities and the accretion of coral reefs: (1) their aggregate structure must produce a depositional environment that retains sediments and propagules; (2) their aggregate production of skeletal material must exceed carbonate losses from that environment; and (3) they must attract secondary producers that bind coral debris, infill framework cavities and recycle nutrients and organic material. Coral communities build reefs, individual corals do not. Coral assemblages in shallow, temperate water often do not form reefs (e.g., zooxanthellate coral communities in SW Australia), while coral assemblages in deep, cold water can build large reefs (e.g., scleractinian mounds off Norway), but do not always do so (e.g., gorgonian assemblages off Newfoundland). Our hypothesis is that the same processes that determine reef accretion in the shallow tropics apply to the deep boreal. An experimental program designed to test this hypothesis involves measures of community structure, bio-particle delivery and depletion, larval settlement, hydrodynamic baffling, sediment composition and deposition along gradients of coral density on the Scotian Slope, where at least twenty species of deep-water corals exist.

INTRODUCTION
Approximately half of the known species of coral lack symbiotic zooxanthellae and live at ocean depths below the photic zone (Cairns and Stanley, 1981). Yet many of them are components of benthic communities that create biogenic structures that may be loosely termed "reefs" (Wilson, 1979; Fagerstrom, 1987). Although the existence of deep coral and the mounds they build has been known for over a century (Verill, 1862; Joubin, 1922; Teichert, 1958), knowledge of them has been limited to descriptions of uplifted fossil bioherms and scattered dredge hauls until the advent of research submersibles. The ecological processes that lead to the origin and growth of deep reefs, and deter-

mine their interactions with adjacent marine communities, remain virtually unknown.

Conversely, the shallow-water reefs built in the tropics by the other half of the world's corals have been the subject of intense study for almost 160 years (Darwin, 1842). Coral reef research is now burgeoning as the threats to these ecologically and economically valuable communities become more obvious and urgent (Hatcher et al., 1989; Wilkinson 2000). As a result, a great deal of empirical data and theoretical modelling exists for shallow, tropical coral reef communities. The recognition of deep coral reefs as biodiversity hotspots, and the continued expansion of deep-sea hydrocarbon production, will lead to a similar burst of research on deep coral communities. It is sensible to attempt to transfer models and lessons learned from shallow coral reef research to deep coral communities.

THE CORAL REEF PARADIGM

Three concepts recur as themes in the well-established field of tropical coral reef science:

a) The well-bounded ecosystem: Coral reefs are consistently modelled and managed as units of organization distinct from adjacent ecosystems (Hatcher, 1997a).
b) The highly-evolved community: Tropical shallow-water coral reefs are the longest surviving, most taxonomically diverse and biocomplex marine benthic biocoenoses on the planet (Wood, 1998).
c) Bio-physical adaptation to the environment: Shallow coral reefs organically respond to and physically-chemically modify the ocean that envelops them (Hubbard, 1997).

We hypothesize that the same processes that determine coral reef integrity, resilience and accretion in the shallow tropics apply to the coral communities of the deep boreal. The hypothesis can be tested by comparing three defining attributes between shallow and deep coral communities, namely:

- Retention, recycling and accumulation of organic and inorganic materials such that the internal biogeochemical fluxes exceed the trans-boundary fluxes. This is a measure of the degree of closure of the reef ecosystem (Hatcher, 1997a).
- Complex biotic interactions within and across trophic levels and functional groups, such that symbiosis, competition and predation exert strong direct and indirect effects on community structure and function (Hatcher, 1997b; Pennings, 1997).
- Homeorhesis at the ecosystem level such that strong feedbacks between biotic and abiotic components serve to maintain community stability up to threshold levels, beyond which massive phase shifts occur (Hatcher, 1984; Hughes, 1994).

Our intent is to first refine and further specify the hypotheses by comparing the known attributes of shallow and deep coral reefs (in this communication). We will then use the results to design and conduct ecological measurements and experiments in deep coral communities aimed at falsifying predictions of system behaviour extrapolated from our knowledge of shallow reef communities.

DEFINITIONS OF TERMS

When comparing deep and shallow coral reefs it is important to clearly define terms. For the purposes of this discussion (and in agreement with common usage, e.g., Fagerstrom, 1987), we recognize that:

Corals capable of forming reefs (bioherms) are hermatypic, foundation species, and may be zooxanthellate or azooxanthellate, scleractinian or gorgonian.

Coral Assemblages are simply groups of co-occurring corals or other taxonomically related organisms within a defined spatial domain; they may not necessarily interact with each other, and they may or may not form communities, reefs or bioherms.

Coral Communities are assemblages of corals and associated organisms that interact to modify their biotic or abiotic environment, but do not necessarily form reefs.

Coral Ecosystems include the coral communities and interacting abiotic components.

Coral Habitat is the environment (including both abiotic and biotic components) in which a coral assemblage occurs. For the purpose of this discussion Shallow coral habitat lies in the photic zone, while Deep coral habitat does not.

Coral Bioherm is the physical manifestation of a coral community's structural history.

Coral Reef is the bioherm plus a surficial veneer that includes both foundation species (e.g., frame-building coral) and associated assemblages.

The distinction between deep and shallow coral reefs is not as clear-cut as the difference between zooxanthellate corals of the photic zone and non-zooxanthellate corals of the aphotic ocean domain (Table 1). Nor is the distinction simply based on the tropical-boreal (latitude) division, depth (shallow/deep) or thermal (warm/cold) regime. For example, non-zooxanthellate

Table 1. Gross habitat requirements of shallow *vs.* deep corals. Optimal and (absolute) ranges of environmental parameters are given. (After Fagerstrom, 1987)

ENVIRONMENT	ZOOXANTHELLATE	NON-ZOOXANTHELLATE
Latitude	23°N–23°S (35°N–32°S)	30°N–70°N (70°N–78°S)
Depth	0–25 m (-0.5–150 m)	60–300 m (-0.5–6200 m)
Temp.	23–28°C (11–40°C)	-10°C (-1–35°C)
Salinity	3–36 (27–48)	34–36 (34–36)
Photic zone	hermatypic & ahermatypic	ahermatypic
Aphotic zone	absent	hermatypic & ahermatypic

corals form Holocene-age reefs as shallow as 60 m depth off Norway (Teichert, 1958), and as deep as 1400 m depth off the Bahamas (Mullins et al., 1981) at approximately the same temperatures (4-9°C), but in intermediate depths (300-600 m) in the Mediterranean at 14°C (Teichert, 1958). Some ahermatypic corals live at depths in excess of 6000 m (Cairns and Stanley, 1981). Predictably, the diversity and growth rates of zooxanthellate corals, and the accretion rates of the reefs they frame, drop sharply with decreasing temperature and increasing depth through the photic zone. The same does not apply to deep corals and the structures they build; factors that are less depth-correlated than light and temperature play the major roles in determining their distribution and abundance (Coates and Kauffman, 1962).

The non-zooxanthellate, hermatypic corals that build deep reefs are heterotrophs that depend on the delivery of appropriate suspended particulate organic matter. They are generally delicate branching and arborescent colonial forms, often with very large polyps (Stetson et al., 1962). It follows that they, like their shallow, zooxanthellate relatives are dependent on the hydrodynamic regime to provide essential nutrition, and susceptible to disturbance by the sedimentation regime (Squires, 1959). Perhaps these corals may be seen as end members of a continuum, at the other end of which symbiosis with zooxanthellae allows corals to thrive in regions where the advective delivery of nutrition is too low to allow their survival as heterotrophs.

ATTRIBUTES OF SHALLOW CORAL REEFS COMPARED TO DEEP REEFS

The defining attribute of a coral reef is that it creates a biogenic depositional environment (Table 2). In shallow reef communities well-differentiated functional groups of constructors, namely framers, cementers and infillers, combine to create 3-D structure, baffle currents, generate and trap sediments, and bind the whole matrix together into a physically robust structure (Hubbard, 1997). A great diversity of calcifying organisms (dominated by, but not exclusively corals) perform these functions in shallow reefs, while a much smaller suite of organisms (again dominated by corals) perform them much more slowly and less completely in deep reefs (Table 2). The result is that deep coral reefs do not attain the size or the permanence in the stratigraphic record that their

Table 2. Biogenic depositional environments defined in terms of the dominant processes in coral reef communities.

STRUCTURE/FUNCTION	SHALLOW REEFS	DEEP REEFS
Vertical accretion	fast (0.5–5 $mm.y^{-1}$)	slow (0.0–0.1 $mm.y^{-1}$)
Framework accretion	dominant	minor
Sediment infilling	significant	dominant
Binding, Bioerosion	essential, rapid	rare, slow (?)
Bioherm size (extent)	1000 m x 50 km (2500 km)	50m x 1 km (1000 km)

shallow counterparts do (Cairns and Stanley, 1981).

Much of the biotic diversity and biomass density of coral reefs may be attributed to the topographic relief and structural complexity of the physical habitat that reefs create (e.g., Roberts and Ormond, 1987; McGehee, 1994, Fig. 1). The multitude of sizes, shapes, exposures and water residence times of cavities, holes, light and current shadows inherent to a coral reef structure provides a great density of habitats across a wide range of spatial scales in which organisms may find shelter, substrate and food (Done et al., 1996). Thus coral reefs have the potential to provide critical habitat to both site-attached and mobile organisms, aggregating them in high densities at reef sites, and thereby facilitating biological interactions like feeding and reproduction. Deep reefs may exhibit densities of corals that are similar to those of shallow reefs (Squires, 1959), but the much lower diversity of structural forms and growth rates of deep corals suggest that the topographic complexity of deep mound reefs (e.g., Freiwald et al., 1997) may be considerably less developed (Table 3). Nonetheless, deep reef structures clearly have the potential to provide aggregating habitat for a wide range of invertebrates and fish in a low-relief environment (Mortenson et al., 1995).

The hydrodynamic regime exerts the dominant control over community structure and function in shallow coral reefs because of its combined effects on trophodynamic fluxes, disturbance and transport of reproductive products (Hamner and Wolanski, 1988). Much of the observed variation in gross morphology (i.e., zonation), nutrient cycling, and recruitment of organisms on reefs can be explained in relation to water movement. Reefs that are hydrodynamically open have stronger connections to adjacent ecosystems and are less self-sufficient than reefs with more closed circulations (Hatcher, 1997a). The slower currents and lack of wave action in deep reef habitats should favour retention and recycling (Table 4), while the low vertical relief and small regenerative space of deep coral mounds may work against hydrodynamic and trophodynamic closure.

Table 3. Topographic and habitat complexity in coral reef communities.

STRUCTURE/FUNCTION	SHALLOW REEFS	DEEP REEFS
Antecedent structure	dominant	weak?
Vertical relief	high (-50 m)	moderate (0.25–5 m)
Fractality	high (~0.8)	unknown (low?)
Zonation	pronounced (5 typical)	unknown (few?)
Rugosity	high (1.5–5)	unknown (low?)
Porosity	high (0.1–500 m.d^{-1})	low?

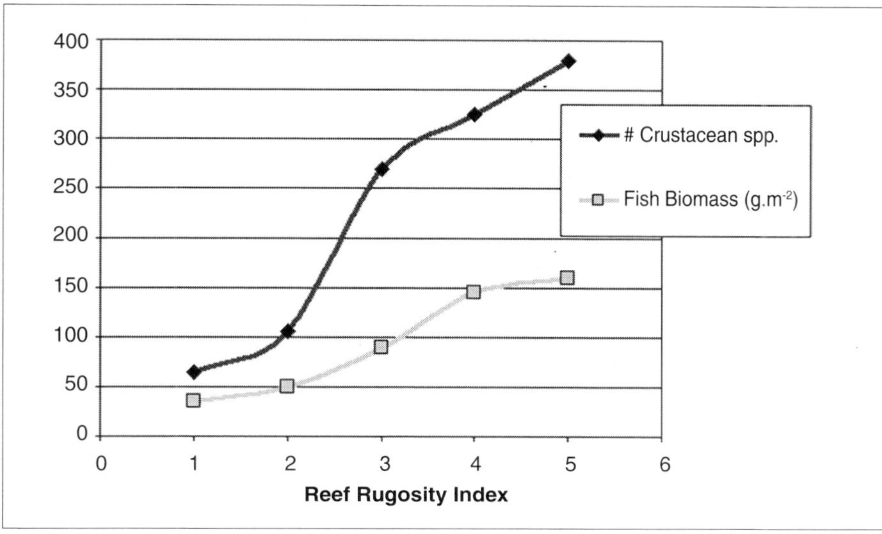

Figure 1. Biodiversity and Biomass of associated organisms increase with coral reef Rugosity (Hatcher, unpublished data, Houtman Abrolhos Reefs, Western Australia)

Concentrated photosynthetic biomass in shallow tropical waters coupled with rapid advection of dissolved nutrients and tight recycling make photic zone coral reefs the most productive communities in the ocean (Hatcher, 1990, 1997b, Fig. 2). Yet only a small fraction of this gross production of organic matter is actually exported from the ecosystem; the rest is recycled such that

Table 4. Indices of the hydrodynamic regime and trophodynamic closure in coral reef ecosystems. Key: U* = near-bed shear Velocity; C_{db} = near-bed Coefficient of drag

STRUCTURE-FUNCTION	SHALLOW REEFS	DEEP REEFS
Advective flux	high (0.2–1.5 $m.s^{-1}$)	high (?)
Exchange time	long (1 d–1 y)	short (?)
U* & C_{db}	high (~0.8, 0.3)	unknown (?)
Symbioses (diversity)	dominant (250 genera)	rare (? general)
Recycling (Finn's Index)	intense (R = 2.2–4.8)	unknown (?)

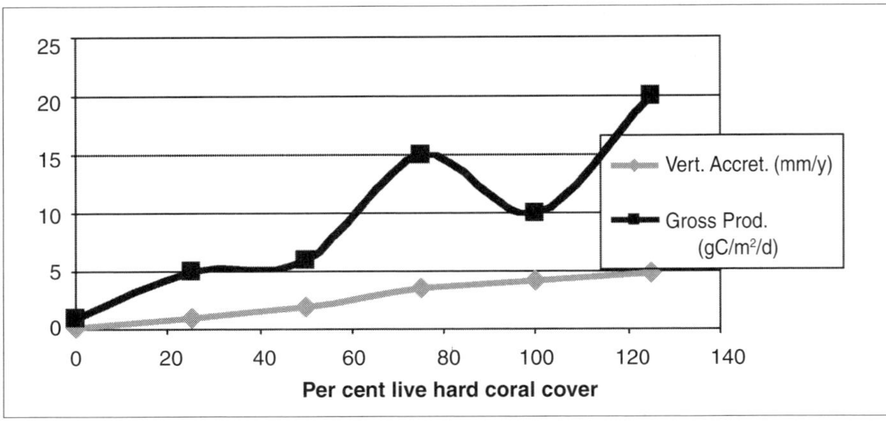

Figure 2. Accretion and Production increase with coral Biomass (data from Hubbard 1997 and Hatcher, 1997)

the net (export) production is little different from that of the plankton communities of the surrounding waters (Crossland et al., 1991). Few comparable measurements of community metabolism are available for deep reefs (Table 5), but the lack of primary producers (except perhaps at sites of chemosynthetic metabolism) and the low rates of organic input to the deep sea mean that deep coral communities will have negative net productivity (i.e., are net heterotrophic). A more significant question is the magnitude of organic

Table 5. Gross, recycled, secondary and excess production on coral reef communities. Key: P=Gross photosynthetic Production; R=Community respiration; E=Excess of P after R over 24 h

STRUCTURE/FUNCTION	SHALLOW REEFS	DEEP REEFS
Photosynthesis (P_{GROSS})	very high (1–20 $gC.m^{-2}.d^{-1}$)	NIL (chemosyn.?) (0.0–0.2 $gC.m^{-2}.d^{-1}$)
Calcification	very high (0.4–10 $kg.m^{-2}.y^{-1}$)	low –moderate?
Organic inputs	low (4–20%)	high (100%)
Inorganic inputs	limiting to E	unlimited
Respiration (Community)	very high (6–15 $gC.m^{-2}.d^{-1}$)	moderate? (0.5–4.0)
Net production (P/R, E)	very low (1.0–1.2), (0.0–0.3 $gC.m^{-2}.d^{-1}$)	low–moderate
Biomass Accumulation	high (10–100 $gC.m^{-2}$) low (10–15%)	low–moderate

Table 6. Biological diversity and functional redundancy of coral reefs.

STRUCTURE-FUNCTION	SHALLOW REEFS	DEEP REEFS
Coral diversity	>700 sp. (mainly Scler.)	> 200 sp.
Species diversity	>93,000 sp.	??? sp.
Endemism	low (<10%)	limited
Life/func. form	variable	limited
Dominance -	low	high
Evenness	mod.-high	low-mod.
Trophic diversity	very high	low
Redundancy	high	low
Reproductive strategies	asexual-colonial sexual-broadcast	asexual-colonial sexual-brooding

input required to meet community respiration, and the relative importance of recycling and accumulation. These are unknowns at present.

Shallow coral reefs have the highest Alpha diversity per unit area of any marine community (Paulay, 1997), for the reasons of long evolution, habitat complexity and disturbance regimes discussed above. The biodiversity is expressed not only at the level of the species, but more impressively at the levels of the functional group, assemblage, community and biogenic habitat. One important aspect of this biodiversity is a considerable degree of redundancy in the biological processes of recruitment, growth, feeding and reproduction required to maintain coral reef functionality (Done et al., 1996). The redundancy is thought to give reefs their demonstrable resiliency through the massive environmental changes in geological time (e.g., glacial-eustacy). Simple contrasts of deep reef communities to this model (Table 6) indicate that their taxonomic and functional diversity is indeed lower (hardly surprising given the of reduced environmental variability of the deep sea), but apparently not so low that functional stability is compromised. Biocomplexity does not necessarily require species richness.

KEY HYPOTHESES

The dynamics of shallow coral reef communities are the most difficult challenge facing reef scientists because of the abundance of strong and weak, direct and indirect interactions, and their variation in space and time. These interactions include symbioses, pathogens, competition, herbivory and predation; all modified and mediated by the abiotic environment and the bio-physical feedbacks (see relevant chapters in Birkeland, 1997). From empirical studies of these interactions, however, have emerged some clear relationships that have predictive power for certain attributes of community structure and function. We use these well-tested hypotheses of shallow reef communities to frame testable hypotheses for deep reef communities.

Hypothesis #1: Deep coral assemblages increase the structural complexity of benthic habitat.

Null Hypothesis:

- Topographic complexity no greater within coral assemblages than on similar substrata outside.

Postulates:

- Rugosity is at least 3x higher within coral assemblages than in coral-free habitats.
- Habitat grain and patch size are at least 3x smaller within coral assemblages.
- Life-form size spectrum is broader within coral assemblages.

Hypothesis #2: Deep coral assemblages alter the hydrodynamic and sedimentary environment.

Null Hypothesis:

- The movement of water and deposition and composition of sediments are no different within a coral community than they are immediately surrounding it.

Postulates:

- Near-bed current velocities are reduced by at least 50%, and turblulence is correspondingly increased within coral assemblages.
- Sedimentation rates are greater and grain size distributions broader in coral assemblages.
- Biogenic sediment layers are much thicker within coral assemblages than outside.
- Biogenic material comprises the bulk of sediments within coral assemblages.

Hypothesis #3: Deep coral assemblages create habitat that supports high biomass, biodiversity and productivity.

Null Hypothesis:

- Standing-stock biomass density of sessile and associated mobile fauna is no higher inside the area of a coral community than it is outside the area of coral occurrence.

Postulates:

- Density of life forms and species richness are at least 10x greater within coral assemblages.
- Living biomass density is at least 10x greater, and organic content of sediments at least 3x greater within coral assemblages.
- Infaunal communities are qualitatively and quantitatively richer within

coral assemblages.

- Abundance of adult, coral-associated organisms is a small fraction of settlers.

Hypothesis #4: Deep coral assemblages are maintained by limited larval dispersal (self-seeding).

Null Hypothesis:

- Deep coral assemblages are dependent on larvae delivered from upstream, and export locally produced reproductive products.

Postulates:

- Settlement of larvae is at least 10x greater within coral assemblages than in adjacent coral-free habitats.
- Populations of corals and associated organisms within coral assemblages include settlers, juveniles and spawners.
- Frequency of lecithotrophic coral larvae is high compared to shallow-water species.

METHODS AND EXPERIMENTAL DESIGN

We intend to begin testing these hypotheses in the summer of 2001 on an oceanographic research cruise to three areas of the Scotian Slope using the Canadian Scientific Submersible Facility: ROPOS (Remotely Operated Platform for Ocean Sciences; http://www.ropos.com). Funding will be provided through an NSERC (National Sciences and Engineering Research Council) grant. This will be the first submersible visit to the putative coral communities that are thought to form along the shelf break and its channels and canyons, where scattered coral colonies have been recovered in dredges and fishing gear (Breeze et al., 1997).

Our research plan calls for the following protocol (see Fig. 3):

- video belt transects (500–1000 m) from beyond coral assemblages into the densest centre for diversity and density measures (stratified random sampling of coral and control areas, n=10)
- vertical and oblique still $1m^2$ photoquadrats for detailed identification of organisms, patch and density analyses (stratified random sampling of coral, boundary and control areas, n=6)
- random-walk video searches at end of each transect to enumerate fish and other coral-associated organisms
- laser-scaled sizing of organisms for size frequency distributions
- box core samples for grain analysis, organic content and infaunal analysis (stratified random sampling of coral, boundary and control areas, n=6)
- samples of corals and associated fauna for identification, and morphological and histological analyses
- sediment traps deployed for 2-4wk periods (stratified random place-

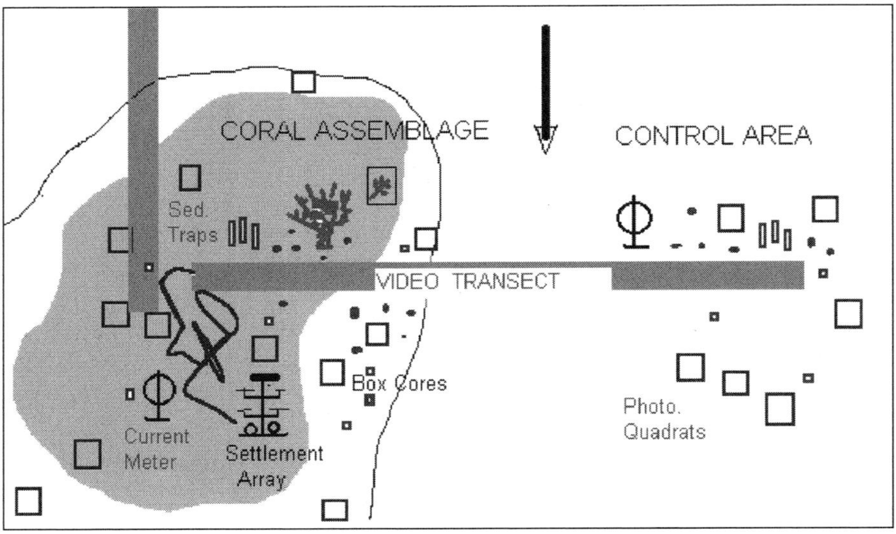

Figure 3. ROV sampling schematic for ecological study of the putative deep coral communities of the Scotian Slope

ment in coral and control areas, n=3 x 3 arrays)

- electromagnetic current and turbulence meters deployed for 2-4wk periods (stratified random placement in coral and control areas, n=2 x 2)
- larval settlement traps and plates deployed for 2-4wk periods (stratified random placement in coral, boundary and control areas, n=3 x 3 arrays)

CONCLUSION

It is time to study the ecology of Canada's deep corals in a holistic sense rather than as taxonomic collections. It remains an open question whether they simply exist as scattered, independent colonies, or actually build reef structures and form ecological communities such as the *Lophelia pertusa* reefs of Norway (Freiwald et al., 1997). The science of shallow tropical coral reefs is becoming mature, and offers robust paradigms and well-formed hypotheses of community structure and function: it is ready for extension out of the shallow tropics (Grigg, 1993). A major commitment of research funds and academic cooperation will be required to extend the current rigour of coral reef science as practised in the sunny tropics to the dark, cold waters of Nova Scotia.

REFERENCES

Birkeland, C.E. (ed.). 1997. Life and death of coral reefs. Chapman and Hall, New York, 536 pp.

Breeze, H., D.S. Davis, M. Butler and V. Kostylev. 1997. Distribution and status of deep sea corals off Nova Scotia. Marine Issues Special Publication No. 1. Ecology Action Centre, Halifax, N.S., 58 pp.

Cairns, S.D. and G.D. Stanley. 1981. Ahermatypic coral banks: living and fossil counterparts, Proc Fourth Intl Coral Reef Symp 57: 611-618.

Coates, A.G. and E.G. Kauffman. 1973. Stratigraphy, paleontology and paleoenvironment of a Cretaceous coral thicket, Lamy, New Mexico, J Paleontol 47: 953-968.

Crossland, C.J., B.G. Hatcher and S.V. Smith.1991. The role of coral reefs in global carbon production, Coral Reefs 10: 55-64.

Darwin, C. 1842. The structure and distribution of coral reefs. Smith, Elder and Co. London, 214 pp.

Done T.J., J.C. Ogden, W.J. Wiebe, D. Bellwood, J. Benzie, C. Birkeland, J. Cortes, C. D'Elia, R. Galzin, P. Glynn, M. Harmelin-Vivien, B.G. Hatcher, J. Jackson, E. Jordan, N. Knowlton, J. McManus, J. Pandolfi, S. Planes, D. Potts, M. Reaka-Kudla, C. Roberts, B. Rosen, P. Sale, B. Salvat, K. Sebens, B. Thomassin, C. Veron and H. Yap (BIOCORE Working Group). 1996. Chapter 15 - Biodiversity and Ecosystem Function of Coral Reefs. In H.A. Mooney, J.H. Cushman, E. Medina, O.E. Sala and E.-D. Schulze (eds), Functional roles of biodiversity: A global perspective. Scientific Committee on Problems in the Environment (SCOPE), J. Wiley and Sons, New York, pp. 393-429.

Fagerstrom J.A. 1987. The evolution of reef communities. J. Wiley and Sons, New York, 600 pp.

Freiwald, A., H. Rudiger and J. Patzold. 1997. Anatomy of a deep-water reef mound from Stjernsund, West Finnmark, Northern Norway. SEPM Special Volume, 56, 141-162.

Grigg, R.W. 1993. Future prospects for coral reef science and species beyond the reef, Coral Reefs 12: 55-56.

Hamner, W.M. and E. Wolanski. 1988. Hydrodynamic forcing functions and biological processes on coral reefs: a status review, Proc Sixth Intl Coral Reef Symp, Townsville, 1: 103-113.

Hatcher, B.G. 1984. A maritime accident provides evidence for alternate stable states in benthic communities on coral reefs, Coral Reefs 3: 199-204.

Hatcher, B.G. 1990. Coral reef primary productivity: A hierarchy of pattern and process, Trends in Ecology and Evolution 5: 149-155.

Hatcher, B.G. 1997a. Coral reef ecosystems: how much greater is the whole than the sum of the parts? Coral Reefs 16(Suppl.): 77-91.

Hatcher, B.G. 1997b. Primary productivity and decomposition. In C. Birkeland (ed.), Life and Death of Coral Reefs, Chapman and Hall, New York, pp. 140-174.

Hatcher, B.G., R.E. Johannes and A.I. Robertson. 1989 Review of research relevant to the conservation of shallow tropical marine ecosystems, Oceanog Mar Biol Ann Rev 27: 337-414.

Hubbard, D.K. 1997. Reefs as dynamic systems. In C.E. Birkeland (ed.), Life and death of coral reefs. Chapman and Hall, New York, pp. 43-67.

Hughes, T.P. 1994. Catastrophes, phase shifts and large-scale degradation of a Caribbean coral reef, Science 265: 1547-1551.

Joubin, L. 1922. Les coraux de mer profunde nuisables aux chalutiers, Notes et Mémoires, Off Sci Tech des Pêche Maritimes 18: 1-16.

McGehee, M.A. 1994. Correspondence between assemblages of coral reef fishes and gradients of water motion, depth, and substrate size off Puerto Rico, Mar Ecol Prog Ser 105: 243-255.

Moretenson, P.B., M. Hovland, T. Brattegard and R. Farestveit. 1995. Deep water bioherms of the Scleractinian coral *Lophelia pertusa* (L.) at 64°N on the Norwegian shelf: structure and associated megafauna, Sarsia 80: 145-158.

Mullins, H.T. 1981. Modern deep water coral mounds north of Little Bahamas Bank: criteria for recognition of deep water coral bioherms in the rock record, J Sediment Petrol 51: 999-1013.

Paulay, G. 1997. Diversity and distribution of reef organisms. In C.E. Birkeland (ed.), Life and Death of Coral Reefs. Chapman and Hall, New York, pp. 298-353.

Pennings, S.C. 1997. Indirect interactions in coral reefs. In C.E. Birkeland (ed.), Life and Death of Coral Reefs. Chapman and Hall, New York, pp. 249-272.

Roberts, C.M. and R.F.G. Ormond. 1987. Habitat complexity and coral reef fish diversity and abundance on Red Sea fringing reefs, Mar Ecol Prog Ser 41: 1-8.

Squires, D.F. 1959. Fossil coral thickets in Wairarapa, New Zealand, J Paleontol 38: 904-915.

Stetson, T.R .1962. Coral banks occurring in deep water on the Blake Plateau, Am Mus Nat Hist Novit 2114, 39 pp.

Teichert, C. 1958. Cold- and deep-water coral banks, Am Assoc Pet Geol Bull 42: 1064-1082.

Verill, A.E. 1862. Notes of a Primnoa from Georges Bank, Proc Essex Inst, Salem MA, 3: 127-129.

Wilkinson, C. (ed.). 2000. The global status of coral reefs. Aust Inst Mar Sci, Townsville, 237 pp.

Wilson, J.B. 1979. Patch development of the deep water coral *Lophelia pertusa* (L.) on Rockall Bank, J Mar Biol Assoc 55: 611-625.

Wood, R. 1998. The ecological evolution of coral reefs, Ann Rev Ecol Syst 29: 179-206.

SOME NOTES ON THE TERMS "DEEP-SEA AHERMATYPIC" AND "AZOOXANTHELLATE," ILLUSTRATED BY THE CORAL GENUS *MADRACIS*

Maya Borel Best
National Museum of Natural History, Naturalis
Leiden, The Netherlands

SUMMARY

The life strategies of scleractinian coral species, and their adaptability to environmental conditions, are heavily discussed topics in coral literature. This has much to do with the presence or absence of symbiotic zooxanthellae in the coral polyp tissue. Deep-sea corals live in the aphotic zone, so are azooxanthellate, and are mostly non-reef-builders, so are ahermatypic. Some deep-sea corals, however, can construct reefs (e.g., *Lophelia*) and are therefore azooxanthellate yet hermatypic. Reef-building corals from shallow waters are in general zooxanthellate, but there are also shallow-water azooxanthellate reef-building corals (e.g., *Dendrophyllia*).

The cosmopolitan coral genus *Madracis*, a member of the family Pocilloporidae, has been recorded from the Mesozoic onward. The genus contains approximately 15 Recent species, having the whole range of life strategies from zooxanthellate-hermatypic to azooxanthellate-ahermatypic. These two extreme states can even be present within one species. Based on examination of systematics, geographical distribution and ecological adaptation, the theory is advanced that some shallow-water (zooxanthellate) *Madracis* species can adapt to deep water, becoming azooxanthellate under certain circumstances, in which case they are called apozooxanthellate, a state that is presumably reversible. This has implications for understanding survival strategies, species variability and distribution patterns.

INTRODUCTION

There is a continuous fossil record of scleractinian corals from the Middle Triassic onward, some 230 million years ago (mya). At that time the world consisted of one supercontinent, Pangea, and one large ocean, the Panthalassa. These corals often had very different ecological preferences from Cenozoic corals; many were solitary and did not form reefs. During the Mesozoic, however, symbiosis with algae furnished evolutionary success for the group (Stanley, 1981). By the Cretaceous, scleractinian corals were forming large reefs in the Tethys Sea, but the catastrophe of 65 mya (presumed to have been caused by a meteorite striking Earth) ended the Cretaceous and led to the extinction of approximately 70% of all coral genera. During the Cenozoic new coral faunae evolved from survivors of the Cretaceous corals, especially from the zooxanthellate members of the family Faviidae.

Some zooxanthellate corals are more adaptable to lower temperature and less light, and can live in deeper habitats than can others. Examples are genera in the Faviidae such as *Favia*, *Diploastrea* and *Hydnophora* (Best, 1974). Another genus of this family is *Cladocora*, a Cretaceous survivor and now the only zooxanthellate in the Mediterranean (Kühlmann, 1996).

Not surprisingly, the azooxanthellate genera were better able to survive the C/T boundary than were their zooxanthellate relatives. Indeed, about half of the azooxanthellate genera survived. The survival rate was particularly high for genera belonging to the family Caryophylliidae. Evolution of coral genera during the Cenozoic was influenced by continental drift (mainly in the Miocene) and sea level changes (in the Plio-Pleistocene).

The family Pocilloporidae has its origin in the Lower Cretaceous, and most genera became extinct in the Lower Tertiary. During the Cenozoic, four major genera (*Stylophora*, *Pocillopora*, *Seriatopora*, *Madracis*) were cosmopolitan, but became restricted to the Tethys Sea during the Miocene. The only survivor of this family in the Atlantic is *Madracis*. This genus has zooxanthellate and azooxanthellate representatives in both the Atlantic and the Indo-Pacific.

Survival of coral species through major geological and climatological changes depends on many factors, but a major reason is the degree to which the animal can adapt to changing environmental conditions.

FACULTATIVE SYMBIOSIS WITH ZOOXANTHELLAE, THE APOZOOXANTHELLATE CONDITION

The highest rates of calcification are found in zooxanthellate corals in warm water (Jacques et al., 1983). Most zooxanthellate corals depend on light to maintain a life strategy of high growth in the highly competitive conditions of a coral reef ecosystem. They are vulnerable, however, in situations when light diminishes (e.g., increase of sedimentation, atmospheric catastrophes).

Some corals seem to have developed a facultative symbiotic life strategy (Fenner, 1993). They can adapt to circumstances when light is scarce or absent by losing their zooxanthellae, yet still continue to grow and reproduce. As stated above, many zooxanthellate coral species, especially those belonging to the Faviidae, can adapt well to decreasing light and lower temperature, assuming an azooxanthellate life without losing their zooxanthellae completely (e.g., *Cladocora*; Kühlmann, 1996).

The phenomenon of facultative symbiosis has been observed in several cases and needs to be studied in more detail in order to understand the underlying mechanism. Examples include species belonging to the ahermatypic genus *Astrangia* (Jacques et al., 1980). Experimental ecological studies have revealed the importance of light, temperature and symbiont density for calcium carbonate production. Another ahermatypic coral genus showing this phenomenon is *Oculina* (Zibrowius, 1980). *Oculina* species have symbionts facultatively and are extremely variable according to depth. The genus is not well studied and is in need of a taxonomic revision (Cairns, 1991).

The only hermatypic genus belonging to the family Pocilloporidae that has species with a facultative symbiotic life strategy is *Madracis*. There are 15 nominal species, some of which show a cosmopolitan distribution pattern. There are enough materials and data available in the Netherlands (RMNH Leiden, ZMA Amsterdam) to be used as basis for an exploratory study, as described below.

SPECIES BELONGING TO THE GENUS *MADRACIS* AND THEIR DISTRIBUTION

In 1999 Cairns presented a list of *Madracis* species and their geographical distribution patterns (Table 1). The aim of the present paper is not a taxonomic revision of the genus, but rather an analysis of the validity of described species in view of the adaptability of coral species to changing environmental conditions. Zooxanthellate corals can adapt more or less to different circumstances, and this well-known, habitat-induced modification of coral species has consequences for taxonomy. But those species that are able to adapt to deep-sea circumstances, losing their zooxanthellate way of life, must have a still greater range of variability. This facultative symbiosis is called the apozooxanthellate condition (for a discussion of the terms zooxanthellate, azooxanthellate and apozooxanthellate, see Schuhmacher and Zibrowius, 1985).

Table 1. *Madracis – Palauastrea* species and their distribution according to Cairns (1999). (wP = west Pacific, eP = east Pacific, wA = west Atlantic, eA = east Atlantic, IO = Indian Ocean, M = Mediterranean)

Species	Condition	Distribution
M. asperula M. Edwards & Haime, 1849	- apozooxanthellate	- wA, eA.
M. hellana M. Edwards & Haime, 1850	- azooxanthellate	- IO
M. kauaiensis Vaughan, 1907	- azooxanthellate	- eP
M. mirabilis sensu Wells, 1973	- zooxanthellate	- wA
M. myriaster (M. Edwards & Haime, 1849)	- azooxanthellate	- wA
M. brueggemanni (Ridley, 1881)	- azooxanthellate	- wA
M. formosa Wells, 1973	- zooxanthellate	- wA
M. decactis (Lyman, 1859)	- zooxanthellate	- wA, eA
M. kirbyi Veron & Pichon, 1976	- zooxanthellate	- wP, eP
M. pharensis (Heller, 1886)	- apozooxanthellate	- wA, eA, M, IO
M. profunda Zibrowius, 1980	- azooxanthellate	- eA
M. senaria Wells, 1974	- zooxanthellate	- wA
M. singularis Rehberg, 1892	- azooxanthellate	- wP
M. interjecta Marenzeller, 1907	- azooxanthellate	- IO
M. asanoi Yabe & Sugiyama, 1936	- azooxanthellate	- wP
? *Palauastrea ramosa* Yabe & Sug., 1941	- zooxanthellate	- wP

A study of the literature and of coral specimens present in collections in the Netherlands (Zoological Museum Amsterdam [ZMA] and National Museum Natural History Leiden [NNM]), revealed the following:

- In the family Pocilloporidae Gray, 1842, five genera are recognized: *Pocillopora*, *Seriatopora*, *Stylophora*, *Palauastrea* and *Madracis*. The first three genera are strictly reef-building and zooxanthellate, the last two both zooxanthellate and azooxanthellate. *Madracis* is the only genus with representatives in the Atlantic and the Pacific, and is the focus of this study.
- The characteristics of the genus *Madracis* Milne Edwards & Haime, 1849, are: colonial, extratentacular budding producing massive or ramose coralla, coenosteum solid, corallites 1-2 mm in diameter, costae in ridges or spines, septa well developed with smooth margins, higher cycles reduced to spines, arranged in 6, 8 or 10, with no more than two cycles, columella styliform, no pali. The type species is *Madracis asperula*.

The following groups of species in the genus *Madracis*, named after the first described species within each group, were identified:

A. Asperula Group (Figures 1a-d)

To this group of *Madracis* corals belong species with the following characteristics in common: 10 equally developed septa, a firm columella and intercalicinal ridges formed by costal spines. These species are found in the west and east Atlantic as well as in the Pacific (i.e., they are cosmopolitan), and at depths ranging from 10 to 70 m. *M. asperula* M. Edwards & Haime, 1849, was first described from Madeira.

The colonies are extremely variable – from encrusting, solid, nodular to branching. The specimens found in shallow water (10-30 m) are solid, encrusting (Figure 1a) to nodular (Figure 1b). The specimens from deeper waters (30-120 m) are all branching colonies with corallites with spiny septa and costal ridges (Figures 1c-d). *M. asperula* appears to be a common species in the western Atlantic (Caribbean and Gulf of Mexico) and the eastern Atlantic from Cape Verdes north to the Madeira Archipelago. In literature it is also mentioned from the Galápagos (Durham and Barnard, 1952; Glynn and Wellington, 1983; Cairns, 1991). Some species described in literature fall within the range of variability of the present Asperula group: *M. hellana*, M. Edwards & Haime, 1850, *M. mirabilis* sensu Wells, both from the West Indies, and *M. kauiensis* Vaughan, 1907, from Hawaii. Cairns (2000: 37) has stated that he could not find any differences between the Atlantic and Pacific populations.

B. Myriaster Group (Figure 2)

The species *M. myriaster* (M. Edwards & Haime, 1849) was redescribed by Cairns in 1979. This is a well-defined species that has been found only in the western Atlantic at depths ranging from 20-700 m. The septal number is 8-20 and the columella is well developed. The most striking characteristics

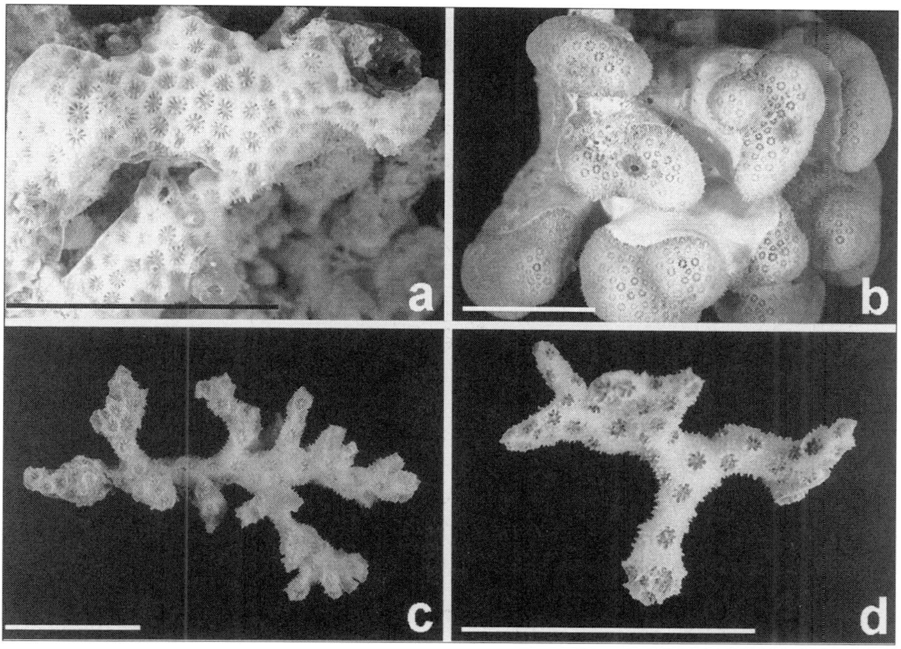

Figure 1. Madracis *"asperula" group (scale bar = 2 cm)*

a. M. asperula *M. Edw. & Haime* *Cape Verdes, 10 m* *RMNH no. 22254*
b. M. asperula *M. Edw. & Haime* *Caribbean, 28 m* *RMNH no. 8606*
c. M. asperula *M. Edw. & Haime* *Cape Verdes, 50 m* *RMNH no. 15972*
d. M. asperula *M. Edw. & Haime* *Suriname, 120 m* *RMNH no. 24260*

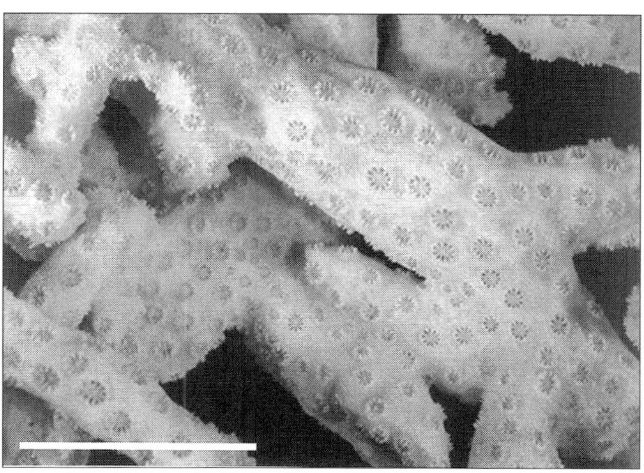

Figure 2. Madracis *"myriaster" group (scale bar = 2 cm)* M. myriaster *(M. Edw. & Haime) Aruba, 60 m RMNH no. 9132*

of this *Madracis* group are the continuous costal striae formed on the coenosteum and the large, thinly or thickly branched (up to 40 cm) bushy colonies. The collection in Leiden contains a large bushy colony (Coel. 9132) in which the large calices (up to 3 mm) contain 8-20 septa.

The species *M. formosa* Wells, 1973, from Jamaica, also falls within the range of variability of the Myriaster group. Thickness of branches is not a specific character. The description of *M. brueggemanni* by Ridley, 1881, demonstrates that this species also falls within the variability of the Myriaster group. The geographical distribution of the group is therefore western Atlantic, from the Caribbean to Brazil.

C. Decactis Group (Figures 3a-e)

The original description of *Astrea decactis* is by Lyman in 1859. The description as well as the figures are concise, but clear enough to recognize a *Madracis* species. *Astrocoenia pharensis*, described by Heller in 1886 as from the Mediterranean, is fully discussed by Zibrowius (1980) as a Mediterranean/eastern Atlantic *Madracis* species. I agree with Fenner's (1993) statement that *pharensis* and *decactis* are probably conspecific, Fenner basing his conclusions on morphological ecological studies. Cairns (2000) describes *M. pharensis* as being highly variable, and raises the possibility that the species may exist as several formae in response to different ecological circumstances.

The main characteristics of this *Madracis* group are: septa hexameral arranged in two cycles, columella are small rods, corallites closely packed and rather large (up to 3 mm). Colonies encrusting to nodular in shallow water and slender nodular branches in deeper water. The later-described species, *M. senaria* Wells, 1974, and *M. profunda* Zibrowius, 1980, fall in the wide variation of the present group as a result of growing in very different circumstances. In 1892 Rehberg described *M. singularis* from the western Pacific. *M. kirbyi* Veron & Pichon, 1976, from Australia was also found in many parts of Indonesia and could not be distinguished from *M. pharensis*. Sheppard & Sheppard (1991) mention the species from the Arabian Gulf. Cairns (2000: 45) writes "a phenotype morphologically indistinguishable from *M. pharensis* has been reported from throughout the Indo-Pacific at 5-421 m."

Figure 3a shows an encrusting *decactis* from shallow water in St. Helena, Figure 3b an encrusting nodular *pharensis* colony from shallow water in Curaçao, Figure 3c a small nodular *pharensis* colony found in a cave near Marseille, Figure 3d a slender branching *profunda* from 160 m depth near the Canary Islands, and Figure 3e a nodular *kirbyi* colony from shallow water at Magnetic Island in Australia. This group of *Madracis* species is cosmopolitan with all adaptations in shallow water due to ecological circumstances; in deep water the species loses its facultatively symbiotic algae. This rather far-reaching statement can only be proven by molecular analysis as proposed by Cairns (2000). Results of genetic research on *Madracis* species in the Caribbean will soon be available (Diekmann et al., 2001).

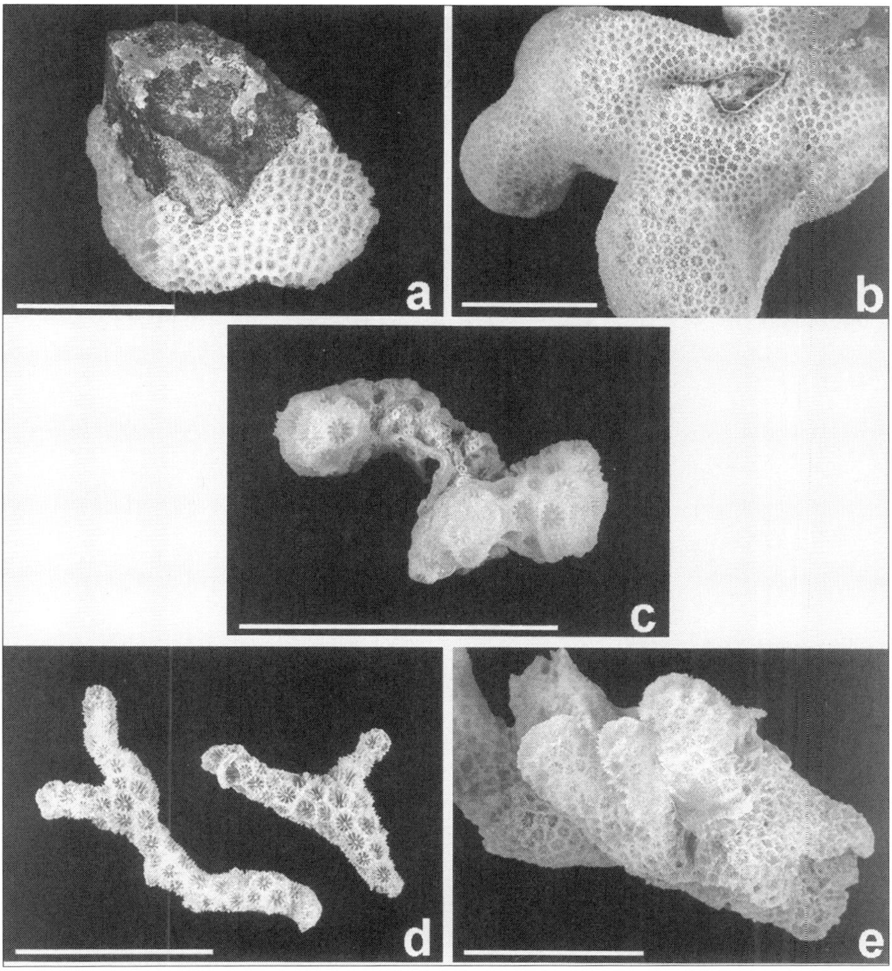

Figure 3. Madracis *"decactis" group (scale bar = 2 cm)*

a. M. decactis *(Lyman)*	*St. Helena, 5 m*	*RMNH no. 18005*
b. M. pharensis *(Heller)*	*Curaçao, 5-10 m*	*RMNH no. 13795*
c. M. pharensis *(Heller)*	*Mediterranean, 10 m*	*ZMA no. 5332*
d. M. profunda *Zibrowius*	*Canary Islands, 160 m*	*RMNH no. 15985*
e. M. kirbyi *Veron & Pichon*	*Great Barrier Reef, 5 m*	*RMNH no. 12214*

D. Interjecta Group (Figure 4)

The species is described by Marenzeller in 1907 as *M. interjecta*. The description and figures are good, but based on a few specimens only. This is a branching species, with small protruding corallites with six well-developed septa, a small rodlike columella and a finely dentate coenosteum. The specimens were found in the Red Sea at depths ranging from 150 to 350 m. Later it was mentioned (again in the Red Sea) by Pillai and Scheer, 1976, and from

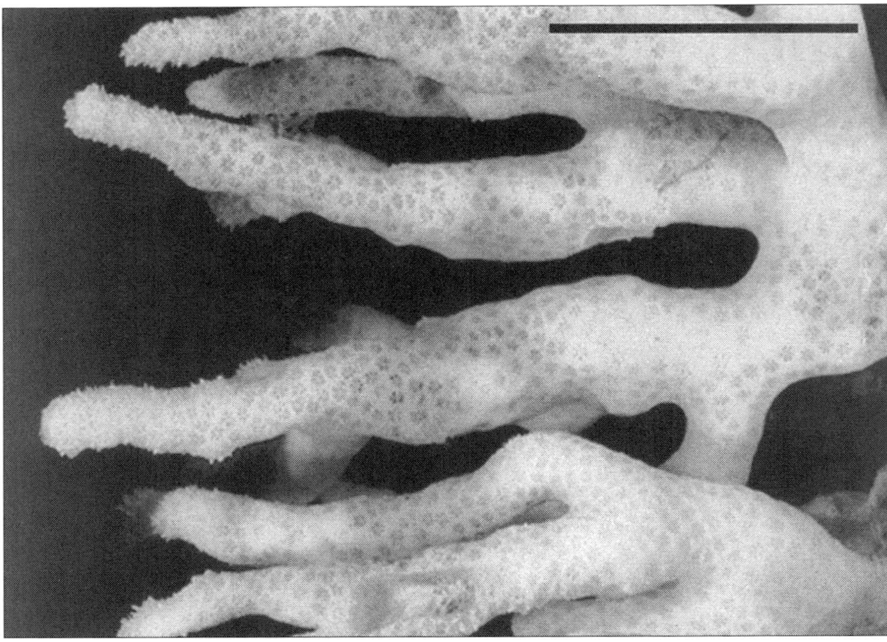

Figure 4. Madracis *"interjecta" group (scale bar = 2 cm)*
Palauastrea ramosa *Yabe & Sug. Sumbawa, 25 m RMNH no. 21629*

the Arabian Gulf by Sheppard and Sheppard in 1991. In 1936 Yabe and Sugiyama described two branching deep-water *Madracis* species from Palau: *M. asanoi* and *M. palaoensis*. The characteristics they have in common are branching coralla and finely dentate coenosteum. The descriptions are very concise.

In the Leiden collections there are specimens from Indonesia from shallow water (Figure 4) identified as *Palauastrea ramosa*. Supposedly azooxanthellate specimens were also found in deeper water (25-50 m) during the *Snellius* expedition. This species, described by Yabe and Sugiyama in 1941, may very well belong to the Interjecta group of branching *Madracis* species, but the type specimen should be studied before any conclusions are reached.

All these species are described from one to a few specimens, so they are rare; we need more material and study in order to know whether we are dealing here with one or more species.

DISCUSSION

The adaptation of sessile zooxanthellate scleractinian coral species to changing environmental physical factors (e.g., depth, light, wave action, sedimentation, temperature) affects colony formation and characteristics of the coral polyps in tropical warm water; this is illustrated by species belonging to the

Faviidae (Best, 1974). Jacques et al. (1983) discuss the interaction of temperature, light and the density of zooxanthellae in the temperate zone and their influence on calcium carbonate production.

The apozooxanthellate coral genus *Madracis* stands out because of the extreme variability of specimens in coral collections, its presence in both the Atlantic and Pacific, and the range of depths (1-750 m) at which it is found. We are dealing here not only with the influence of physical factors in shallow tropical water, but also with the adaptation by members of the genus to deep and cold water by means of the facultative presence of zooxanthellae.

For *Madracis* species it has been possible to survive in colder water and to migrate to deeper water. This explains the wide geographical distribution of most species and the survival of the genus in the Atlantic after the closure of the Panama zone in the Pliocene.

The four groups of *Madracis* distinguished are not taxonomic units, but it is quite possible that most of the described Recent species are mere growth forms.

- The Asperula group comprises apozooxanthellate species or growth forms (*asperula*, *hellana*, *mirabilis*, *kauiensis*) and is cosmopolitan.
- The Myriaster group comprises apozooxanthellate species or growth forms (*myriaster*, *formosa*) occurring in the west Atlantic.
- The Decactis group comprises apozooxanthellate species or growth forms (*decactis*, *pharensis*, *profunda*, *kirbyi*, *senaria*, *singularis*) and is cosmopolitan.
- The Interjecta group comprises apozooxanthellate species or growth forms (*interjecta*, *asanoi*, *? P. ramosa*) occurring in the Indian and Pacific Ocean.

CONCLUSIONS

Corals belonging to the genus *Madracis* include both zooxanthellate and azooxanthellate species, and occur in many different biotopes, from shallow coastal tropical water to temperate deep-water areas. Some species are restricted to particular ecological circumstances; others (by way of an as yet unknown mechanism) can adapt to extremely different circumstances of temperature and depth as a result of facultative symbiosis of their zooxanthellae. These apozooxanthellate coral species are therefore highly variable as far as their morphological characteristics are concerned. In view of this, the taxonomic status of the 15 described Recent *Madracis* species should be reconsidered.

ACKNOWLEDGEMENTS

I would like to thank Mr. A. 't Hooft and Mr. E.J. Bosch for making the illustrations, Prof. G.J. Boekschoten and Dr. S.D. Cairns for their critical reading of the text.

REFERENCES

Best, M.W. 1974. Habitat induced modification of reef building corals (Faviidae) and its consequences for taxonomy. Proceedings of the Second Coral Reef Symposium. 2. Great Barrier Reef Commission, pp. 217-228.

Cairns, S.D. 1991. A revision of the ahermatypic Scleractinia of the Galápagos and Cocos Islands, Smithsonian Contributions to Zoology 504: 1-32.

Cairns, S.D. 1999. Species richness of Recent Scleractinia, Atoll Research Bulletin 459: 1-46.

Cairns, S.D. 2000. A revision of the shallow-water azooxanthellate Scleractinia of the western Atlantic, Studies Natural History Caribbean 75.

Diekmann, O.E., R.P.M. Bak, W.T. Stam and J.L. Olsen. 2001. Molecular genetic evidence for probable reticulate speciation in the coral genus *Madracis* from a Caribbean fringing reef slope, Marine Biology (in press).

Durham, J.W. and J.L. Barnard. 1952. Stony corals of the Eastern Pacific collected by the Velero III and Velero IV, Alan Hancock Pacific Expedition 16(1): 1-110.

Fenner, D.P. 1993. Species distinctions among several Caribbean stony corals, Bulletin of Marine Science 53(3): 1099-1116.

Glynn P.W. and G.M. Wellington. 1983. Corals and coral reefs of the Galápagos Islands. University of California Press, Berkeley, 330 pp.

Heller, C. 1868. Die Zoophyten und Echinodermen des Adriatischen Meeres. Wien, pp. 1-88.

Jacques, T.G. and M.E.Q. Pilson. 1980. Experimental ecology of the temperate Scleractinian coral *Astrangia danae*. I. Partition of respiration, photosynthesis and calcification between hosts and symbionts, Marine Biology 60: 167-178.

Jacques, T.G., N. Marshall and M.E.Q. Pilson. 1983. Experimental ecology of the temperate Scleractinian coral *Astrangia danae*. II. Effect of temperature, light intensity and symbiosis with zooxanthellae on metabolic rate and calcification, Marine Biology 76: 135-148.

Kühlmann, D.H.H. 1996. Preliminary report on Holocene submarine accumulations of *Cladocora caespitosa* (L. 1767) in the Mediterranean. In J. Reitner, F. Neuweiler and F. Gunkel (eds), Global and regional controls on biogenic sedimentation. I. Reef evolution. Research reports. Göttiger Arbeiten Geologie Paläontogie 2: 65-69.

Lyman, H. 1859. On a new species of coral, Proceedings of the Boston Society of Natural History 6: 260-263.

Marenzeller, E.v. 1907. Expeditionen S.M. Schiff Pola in das Rote Meer, nörtliche und südliche Hälfte 1895/96 – 1897/98. Zoologischen Ergebnissen, Tiefseekorallen 25: 13-25.

Milne Edwards, H. and J.M. Haime. 1849. Mémoire sur les polypiers appartenant à la famille des Oculinides, au groupe intermédiaire des pseudoastérides et à la famille des Fongides, Academie de la Science, Paris C.R. 29: 67-73.

Milne Edwards, H. and J.M. Haime. 1850. Recherches sur les polypiers. Cinqième mémoire. Monographie des Oculinides, Annales Sciences Naturalles, Zoologie 3(10): 63-110.

Pillai, G.S., Gopinadha and G. Scheer. 1976. Report on the stony corals from the Maldive archipelago. Results of the Xarifa expedition 1957/58 of the International Institute for Submarine Research, Vaduz, Liechtenstein (Director Dr. Hans Hass), Zoologica 126: i-iv, 1-83.

Rehberg, H. 1892. Neue und wenig bekannte Korallen, Abhandlungen aus dem Gebiete der Naturwissenschaften 12(1): 1-50.

Ridley, S.O. 1881. Account of the zoological collections made during the survey of H.M.S. Alert in the Straits of Magellan and on the coast of Patagonia. X. Coelenterata, Proceedings Zoological Society, London, pp. 101–107.

Schuhmacher, H. and H. Zibrowius. 1985. What is hermatypic? A redefinition of ecological groups in corals and other organisms, Coral Reefs 4(1): 1-9.

Sheppard, C.R.C. and A.L.S. Sheppard. 1991. Corals and Coral Communities of Arabia. Fauna of Saudi Arabia 12: 3-170.

Stanley, G.D. 1981. Early history of scleractinian corals and its geological consequences, Geology 9: 507-511.

Vaughan, T.W. 1907. Recent Madreporaria of the Hawaiian Islands and Laysan, Bulletin United States National Museum 59: i-iv, 4-427.

Veron, J.E.N. and M. Pichon. 1976. Scleractinia of eastern Australia. I. Fam. Thamnasteriidae, Astrocoeniidae, Pocilloporiidae. Australian Institute Marine Science, Monograph Series 1: 1-86.

Wells, J. W. 1973. New and old Scleractinia from Jamaica, Bulletin Marine Science 23(1): 16-58.

Wells, J.W. 1973b. Two new hermatypic scleractinian corals from the West Indies, Bulletin Marine Science 23(4): 925-932.

Yabe, H. and T. Sugiyama. 1936. Some deep-water corals from the Palao Islands, Proceedings Imperial Academy Japan 12: 346-349.

Yabe, H. and T. Sugiyama. 1941. Recent reef-building corals from Japan and the South Sea Islands under the Japanese mandate. Pt. 2. Science Reports. Tôhoku Imperial University. ser. 2, Geology, special volume 2: 67-91.

Zibrowius, H. 1980. Les scléractiniaires de la Méditerranée et de l'Atlantique nord-oriental, Mémoires de l'Institut Océanographique, Monaco 11: 1-284.

BIOGEOGRAPHIC AFFINITIES OF THE NORTH ATLANTIC DEEP-WATER SCLERACTINIA

Stephen D. Cairns and Ralph E. Chapman
National Museum of Natural History, Smithsonian Institution
Washington, D.C., USA

SUMMARY

We investigated the biogeographic properties and relationships of Scleractinia in the North Atlantic. The North Atlantic was divided into 38 geographic regions and each of the 134 North Atlantic species known from deeper than 200 m was scored for each region, resulting in a data matrix of 5092 cells. The species composition of each region was then analyzed by calculating a matrix of Bray-Curtis distance coefficients and performing a cluster analysis (UPGMA) and ordination (Non-Metric Multi-Dimensional Scaling) to determine biogeographic affinities and relationships of these localities. Three major clusters (superclusters) were distinguished: western, eastern and North Atlantic. The western Atlantic supercluster consists of the tropical and warm temperate western Atlantic regions and is characterized by 58% endemic species and a relatively low percentage of deep-water species. It consists of three subordinate clusters, which represent insular, continental and warm temperate areas. The eastern Atlantic supercluster extends from the Faroe Islands to the Gulf of Guinea, and is characterized by 38% endemic species and a disproportionately high number of amphi-Atlantic and cosmopolitan species, as well as a higher than average number of deep-living species. It also consists of three subordinate clusters: a Lusitanian warm temperate group, tropical localities, and a third cluster consisting of the Cape Verde and Canary Islands. The North Atlantic supercluster is characterized not by endemic species, but rather by a high number of cosmopolitan and amphi-Atlantic species, as well as a disproportionately high number of deep-living, unattached, and flabellid species. Two subordinate clusters correspond to the cold temperate northeast and northwest Atlantic.

INTRODUCTION

There is a deep-rooted desire among people to organize data into systems of classification (e.g., clusters). This probably originates in the need to organize multi-dimensional data into smaller packets that can be visualized and manipulated by researchers, but also probably reflects a degree of real clustering in natural systems. Within natural history, if species are the objects of classification, then cladistics or phenetics may provide the algorithms used, but if zoogeography is the subject under consideration, then geographic regions are usually the units of classification and various clustering and ordination algorithms are the tools typically used to synthesize these data. According to Clarke and Warwick (1994) there are literally hundreds of clustering methods, indeed whole volumes (e.g., Hartigan, 1975) dedicated just to clustering algorithms. Cormack (1971) warned that the "availability of ... classification techniques has led to the waste of more valuable scientific time than any other 'statistical' innovation." Nonetheless, a judicious use of clustering and ordination techniques can lead to thought-provoking and valuable insights about the distribution of organisms and it is becoming increasingly common for authors of biogeographic analyses to include a clustering dendrogram and/or an ordination to explore their data set. Several examples using cnidarian taxa are presented below, in increasing order of sophistication, each with a slightly different method and purpose for the analyses, but all using the Bray-Curtis or Jaccard similarity coefficients and various clustering (typically UPGMA; Sneath and Sokal, 1973) and ordination algorithms (usually non-metric multi-dimensional scaling, MDS; Clarke and Warwick, 1994).

For instance, Veron (1995) used clustering techniques in a simple way to document patterns of geographic affinity of shallow-water Scleractinia from adjacent geographic regions. Similarly, Wallace (1999) used clustering methods to demonstrate geographic affinities among acroporid Scleractinia, but also analyzed the composition of her clusters based on the overall distribution of the component species. Ketchum and Reyes-Bonilla (1997) and Reyes-Bonilla and López-Pérez (1998), based on shallow-water Scleractinia, clustered the regions of the eastern Pacific followed by an analysis of these patterns for evidence of the presumed direction of colonization, and to test for the validity of zoogeographic barriers. In an elegant pair of papers based on shallow-water corals, Sheppard (1997, 1999) used clustering and ordination (MDS) to find the geographic affinities of various localities in the Indian Ocean and then used these results to assess the role of the Chagos Archipelago as a possible stepping stone for dispersal. Clustering and ordination can also be used for studying finer-scale ecological processes, such as assessing the relationship between a circumscribed ecological zone of occurrence versus various physical factors, such as light and substrate, as was done by Sánchez et al. (1998) and Sánchez (1999) for shallow-water Caribbean Octocorallia on selected reefs.

The current study seeks to use clustering and ordination analyses to address many of the parameters described above. First, we seek to establish how many species of Scleractinia occur in the North Atlantic below 200 m, where they occur, and at what depths. Using these data, we then try to document the relationships among the faunas of various regions, in both gross and fine detail, based on the shared presence of these taxa. We attempt to characterize these groupings based on physical characteristics of the environment and the morphological characteristics of the species, in an effort to understand why these various localities cluster in the way they do. In so doing we compare our clusters to the traditional shallow-water zoogeographic provinces. We also compare the results of traditional clustering with the results of the ordination (MDS) in an effort to better understand the clusters. Finally, we ask if those transitional localities that do not fit nicely within any single cluster may, instead, represent either stepping stones for dispersal or are simply regions for which little is known.

MATERIAL AND METHODS

Material

The taxonomy and distribution of the North Atlantic deep-water Scleractinia are fairly well known, and are based primarily on the faunistic revisions of the western Atlantic (Cairns, 1979) and eastern Atlantic (Zibrowius, 1980). These sources were updated with records from Zibrowius (1983, 1985, 1992), Cairns (1981, 2000), and Tyler and Zibrowius (1992). From these sources, a total of 134 species (Table 1) are known to occur at depths greater than 200 m in the North Atlantic, representing about 20% of the worldwide azooxanthellate fauna (Cairns, 1999). For the purpose of these analyses, the two subspecies of *Caryophyllia ambrosia* are treated as different species. Drawing on these 134 species, species lists were compiled for 38 regions (Table 2, Figure 1) in the North Atlantic, in general corresponding to those regions listed by Cairns (1979) and Zibrowius (1980) in their respective analyses; in our study, however, the additional outlying locality of Indo-Pacific was included. The only regions not covered in the analysis were the coastline of western Africa between Guinea and Ivory Coast and the New England Seamounts, areas that are still very poorly known.

Methods

The first step in any biogeographic analysis of this type is to examine the data matrix for endemic species, i.e., taxa that are restricted to a single locality in our data set. Although quite useful for documenting the diversity present in each locality, such species are not useful for showing the relationship among the various localities. Consequently, they were not used for the generation of the distance matrix used for the clustering and ordinations; fifteen taxa were noted as endemic and were removed from the data matrix, leaving a 38 x 119 matrix. We then performed an outlier analysis using the program PC-ORD4 (McCune and Mefford, 1999) to see which localities are so different from the others that they should also be removed so as not to overly affect the analyses. Two localities, St. Peter and Paul Rocks and the Arctic (Table 2), were determined to be outliers and were removed from the subsequent analyses, although they subsequently were used for the overall interpretation of the data. In preliminary cluster analyses, these two localities always joined the rest at the very end, independent from each other. The result was the 36 x 119 data matrix used for the clustering and ordination analyses (Table 3).

The next step in the process is the generation of the distance (0 = dissimilarity; 1 = similarity) matrix that serves as the starting point for the more advanced analyses. There are a great many coefficients that could be used including euclidean distance, correlation coefficients and other angle measurements, and coefficients of association such as the Jaccard or Dice (Sneath and

Sokal, 1973; Clarke and Warwick, 1994; McCune and Mefford, 1999). We chose to apply the Bray-Curtis Coefficient (also known as the Sorensen or Czekanowski Coefficient; Bray and Curtis, 1957; Clarke and Warwick, 1994; McCune and Mefford, 1999), which is related to the city-block metric distance measure (Sneath and Sokal, 1973; McCune and Mefford, 1999). It has become the most commonly applied coefficient for studies in quantitative biogeographic and ecological analyses and their paleontological equivalents because it typically exhibits a little more sensitivity to trends when compared with euclidean distance. Furthermore, it is not affected by joint absences of taxa; gives more weight to abundant species than rare ones; has some theoretical support based on fuzzy-set theory (Roberts, 1986); and empirically tends to give very useful results (Clarke and Warwick, 1994; McCune and Mefford, 1999).

The Bray-Curtis distance coefficient is relatively simple to calculate for presence/absence data such as we used. The numerator is the sum of the absolute value of the differences between the equivalent entries for the two localities being compared. If a species is either present or absent in both localities, then the entry for the species is 0. If a species is present in only one locality, the entry is 1. These entries are then summed across all species to give the numerator value. The denominator is the sum of all entries for both localities, equivalent to the total number of occurrences of species at both localities. If the two localities are exactly mis-matched, the numerator equals the denominator and the distance is the maximum value of one. If the localities are exactly matched, the numerator, and thus the coefficient, equals zero; the two localities have no distance between them in space.

The final result was a 36 x 36 matrix of distance coefficients for each locality with itself and each of the other localities. The distance of a locality with itself is, of course, zero; these form the diagonal elements of the matrix. The other two sections (top and bottom) of the matrix are then symmetrical as the distance of locality 1 with locality 2 equals the distance of locality 2 with locality 1. This matrix was then used with the selected clustering algorithm to produce the dendrogram. UPGMA (Unweighted Pair-Group Method with Arithmetic Averages; Sneath and Sokal, 1973) was chosen because it has been shown to be the algorithm that produces the best clustering representation of the original distance matrix using the cophenetic correlation coefficient as an indicator (Sneath and Sokal, 1973) and is, by far, the most commonly applied clustering algorithm in natural history applications. Once the clusters were established (Figure 2, Table 4), an effort was made to characterize each cluster by annotating various biological data exhibited by the localities that make up that cluster. These include the type of distribution patterns shared by the species found at those localities, the number of deep-living species, the percentage of colonial and attached species, and the ratio of the major taxonomic groups.

However useful clustering techniques may be in establishing and helping to visualize the relationships of the entities being studied (i.e., regions), clustering also has several disadvantages that have been discussed by various authors including Sneath and Sokal (1973) and Field et al. (1982). These include

Table 1. Taxa used in the analyses. *Number*: # assigned in data matrix (Table 3); those species not assigned a number are endemic to one region, as indicated in column 2 (for abbreviations see Table 2). *Pattern*: pattern of distribution (or endemicity): w – endemic to western Atlantic; e – endemic to eastern Atlantic; aad – amphi-Atlantic with a disjunct distribution; aac – amphi-Atlantic with contiguous distribution; c – widespread (cosmopolitan) distribution. *Coloniality*: colonial (C) or solitary (S). *Attachment*: attached (A) or free-living (F). C/F/D: caryophylliid, flabellid or dendrophylliid. *Depth*: depth range in western Atlantic (w) and eastern Atlantic (e).

Number	Endemic	Taxon	Pattern	Coloniality	Attachment	C:F:D	Depth (m)
1		*Anomocora fecunda* (Pourtalès, 1871)	aad	C	F	C	w:37-640; e:130-350
2		*Anomocora marchadi* (Chevalier, 1966)	c	S	F	C	w:35-229; e:32-85
3		*Anomocora prolifera* (Pourtalès, 1871)	aad	S	F	C	w:30-329; e:110-125
4		*Anthemiphyllia patera patera* Pourtalès, 1878	w	S	F	-	w:500-700
5		*Astrangia poculata* (Ellis & Solander, 1786)	w	C	A	-	w:0-263
6		*Aulocyathus atlanticus* Zibrowius, 1980	e	S	A	C	e:450-1716
	NWC	*Balanophyllia bayeri* Cairns, 1979	w	S	A	D	w:274-311
7		*Balanophyllia cellulosa* Duncan, 1873	e	S	A	D	e:80-850
8		*Balanophyllia cyathoides* (Pourtalès, 1871)	w	S	A	D	w:45-494
9		*Balanophyllia dineta* Cairns, 1977	w	S	A	D	w:127-274
10		*Balanophyllia floridana* Pourtalès, 1868	aad	S	A	D	w:13-220; e:29-95
	SWC	*Balanophyllia hadros* Cairns, 1979	w	S	A	D	w:238-274
11		*Balanophyllia palifera* Pourtalès, 1878	w	S	A	D	w:53-708
12		*Balanophyllia thalassae* Zibrowius, 1980	e	S	A	D	e:380-1150
13		*Balanophyllia wellsi* Cairns, 1977	w	S	A	D	w:412-505
14		*Bathypsammia fallosocialis* Squires, 1959	w	S	A	D	w:213-805
15		*Bathypsammia tintinnabulum* (Pourtalès, 1868)	w	S	A	D	w:210-1115
16		*Caryophyllia abyssorum* Duncan, 1873	e	S	A	C	e:732-2000
	AZO	*Caryophyllia alberti* Zibrowius, 1980	e	S	A	C	e:76-506
17		*Caryophyllia ambrosia ambrosia* Alcock, 1898	c	S	F	C	w:1487-2286; e:1600-2300
18		*Caryophyllia ambrosia caribbeana* Cairns, 1979	w	S	F	C	w:183-1646
19		*Caryophyllia antillarum* Pourtalès, 1874	w	S	A	C	w:150-730
20		*Caryophyllia atlantica* (Duncan, 1873)	e	S	A	C	e:1100-2165
21		*Caryophyllia barbadensis* Cairns, 1979	w	S	A	C	w:129-249
22		*Caryophyllia berteriana* Duchassaing, 1850	w	S	A	C	w:99-1033
23		*Caryophyllia calveri* Duncan, 1873	e	S	A	C	e:91-340
24		*Caryophyllia corrugata* Cairns, 1979	w	S	A	C	w:183-380
25		*Caryophyllia cyathus* (Ellis & Solander, 1786)	e	S	A	C	e:70-300
	AZO	*Caryophyllia foresti* Zibrowius, 1980	e	S	A	C	e:155-950
	LA	*Caryophyllia paucipalata* Moseley, 1881	w	S	A	C	w:714-843
26		*Caryophyllia polygona* Pourtalès, 1878	w	S	A	C	w:700-1817
27		*Caryophyllia sarsiae* Zibrowius, 1974	aad	S	A	C	w:1097-1200; e:520-2200

Table 1. Taxa used in the analyses *(continued)*

Number	Endemic	Taxon	Pattern	Coloniality	Attachment	C:F:D	Depth (m)
28		*Caryophyllia seguenzae* Duncan, 1873	e	S	F	C	e:1000-1400
29		*Caryophyllia smithii* Stokes & Broderip, 1828	e	S	A	C	e:40-400
30		*Caryophyllia zopyros* Cairns, 1979	w	S	A	C	w:73-618
31		*Ceratotrochus magnaghii* Cecchini, 1914	e	S	A	C	e:7-400
32		*Cladocora debilis* ME & H, 1849	aad	C	F	C	w:32-480; e:28-100
33		*Cladopsammia manuelensis* (Chevalier, 1966)	aad	C	A	D	w:70-366; e:55-150
34		*Coenocyathus anthophyllites* ME & H, 1848	e	C	A	C	e:65-250
35		*Coenocyathus cylindricus* ME & H, 1848	e	C	A	C	e:65-500
36		*Coenocyathus parvulus* (Cairns, 1979)	w	C	A	C	w:97-399
37		*Coenosmilia arbuscula* Pourtalès, 1874	w	C	A	C	w:74-622
38		*Colangia immersa* Pourtalès, 1871	w	C	A	C	w:0.5-347
39		*Concentrotheca laevigata* (Pourtalès, 1871)	aad	S	A	C	w:183-576; e:600-772
	NWC	*Crispatotrochus* sp cf *C. cornu* Moseley, 1881	w	S	A	C	w:220-241
40		*Crispatotrochus squiresi* Cairns, 1979	w	S	A	C	w:686-822
	GSN	*Cryptotrochus carolinensis* Cairns, 1988	w	S	F	-	w:320-338
41		*Dasmosmilia lymani* (Pourtalès, 1871)	c	S	F	C	w:37-366; e:85-316
42		*Dasmosmilia variegata* (Pourtalès, 1871)	c	S	F	C	w:110-421; e:185-600
43		*Deltocyathoides stimpsonii* (Pourtalès, 1871)	aad	S	F	-	w:110-553; e:200-600
44		*Deltocyathus agassizii* Pourtalès, 1867	w	S	F	C	w:494-1115
45		*Deltocyathus calcar* Pourtalès, 1874	w	S	F	C	w:81-675
46		*Deltocyathus eccentricus* Cairns, 1979	aad	S	F	C	w:183-907; e:300-1000
47		*Deltocyathus italicus* (Michelotti, 1838)	aad	S	F	C	w:403-2634; e:1500-2300
48		*Deltocyathus moseleyi* Cairns, 1979	aad	S	F	C	w:201-777; e:200-1372
49		*Deltocyathus pourtalesi* Cairns, 1979	w	S	F	C	w:311-567
50		*Dendrophyllia alternata* Pourtalès, 1880	aad	C	A	D	w:276-900; e:450-688
51		*Dendrophyllia cornigera* (Lamarck, 1816)	e	C	A	D	e:80-600
52		*Desmophyllum dianthus* (Esper, 1794)	c	S	A	C	w:183-2250; e:230-2460
53		*Desmophyllum striatum* Cairns, 1979	w	S	A	C	w:277-823
54		*Eguchipsammia cornucopia* (Pourtalès, 1871)	aad	C	F	D	w:91-300; e:330-960
55		*Eguchipsammia gaditana* (Duncan, 1873)	c	C	F	D	w:97-505; e:73-417
56		*Enallopsammia profunda* (Pourtalès, 1867)	w	C	A	D	w:403-1748
57		*Enallopsammia rostrata* (Pourtalès, 1878)	c	C	A	D	w:300-1646; e:480-2165

Table 1. Taxa used in the analyses *(continued)*

Number	Endemic	Taxon	Pattern	Coloniality	Attachment	C:F:D	Depth (m)
58		*Flabellum alabastrum* Moseley, 1873	aac	S	F	F	w:357-1977; e:906-2000
59		*Flabellum angulare* Moseley, 1876	aac	S	F	F	w:2266-3186; e:2000-2700
60		*Flabellum atlanticum* Cairns, 1979	w	S	F	F	w:357-618
61		*Flabellum chunii* Marenzeller, 1904	e	S	F	F	e:200-700
62		*Flabellum floridanum* Cairns, 1991	w	S	F	F	e:80-366
63		*Flabellum macandrewi* Gray, 1849	aac	S	F	F	w:180-667; e:128-1260
64		*Flabellum moseleyi* Pourtalès, 1880	w	S	F	F	w:216-1097
65		*Fungiacyathus crispus* (Pourtalès, 1871)	aad	S	F	-	w:183-1115; e:340-1010
66		*Fungiacyathus fragilis* Sars, 1872	c	S	F	-	w:412-460; e:200-2200
67		*Fungiacyathus marenzelleri* (Vaughan, 1906)	c	S	F	-	w:2450-4798; e:1820-4822
68		*Fungiacyathus pusillus* (Pourtalès, 1868)	w	S	F	-	w:285-685
69		*Fungiacyathus symmetricus* (Pourtalès, 1871)	w	S	F	-	w:183-1664
70		*Gardineria paradoxa* (Pourtalès, 1868)	c	S	A	-	w:91-700
71		*Gardineria simplex* (Pourtalès, 1878)	w	S	A	-	w:46-241
72		*Guynia annulata* Duncan, 1872	c	S	A	-	w:30-653; e:28-200
73		*Javania cailleti* (Duch. & Mich., 1864)	c	S	A	F	w:30-1809; e:400-2165
74		*Javania pseudoalabastra* Zibrowius, 1974	aad	S	A	F	w:625-1234; e:1070-1275
75		*Labyrinthocyathus facetus* Cairns, 1979	w	S	A	C	w:385-402
76		*Labyrinthocyathus langae* Cairns, 1979	w	S	A	C	w:695-810
77		*Leptopenus discus* Moseley, 1881	c	S	F	-	w:2842-3475
	SHE	*Leptopsammia britannica* (Duncan, 1870)	e	S	A	D	e:459-665
	MAU	*Leptopsammia chevalieri* Zibrowius, 1980	e	S	A	D	e:70-500
	AZO	*Leptopsammia formosa* Gravier, 1915	e	S	A	D	e:375-1385
78		*Lophelia pertusa* (L, 1758)	c	C	A	C	w:146-1200; e:50-1440
79		*Madracis asperula* ME & H, 1849	c	C	A	-	w:24-311; e:0-95
80		*Madracis myriaster* (ME & H, 1849)	w	C	A	-	w:37-1220
81		*Madracis pharensis pharensis* (Heller, 1868)	c	C	A	-	w:11-333; e:6-40
82		*Madracis profunda* Zibrowius, 1980	e	C	A	-	e:112-327
83		*Madrepora carolina* (Pourtalès, 1871)	w	C	A	-	w:53-801
84		*Madrepora oculata* L, 1758	c	C	A	-	w:300-1391; e:295-1550
85		*Oxysmilia rotundifolia* (ME & H, 1848)	w	S	A	C	w:46-640
86		*Paracyathus arcuatus* Lindström, 1877	e	S	A	C	e:201-343
87		*Paracyathus pulchellus* (Philippi, 1842)	aad	S	A	C	w:17-250; e:40-1260

Table 1. Taxa used in the analyses *(continued)*

Number	Endemic	Taxon	Pattern	Coloniality	Attachment	C:F:D	Depth (m)
88		*Peponocyathus folliculus* (Pourtalès, 1868)	aad	S	F	-	w:284-457; e:300-732
89		*Phacelocyathus flos* (Pourtalès, 1878)	w	C	A	C	w:20-355
90		*Placotrochides frustum* Cairns, 1979	aad	S	F	F	w:497-907; e:1300
91		*Polycyathus mayae* Cairns, 2000	w	C	A	C	w:137-309
92		*Polymyces fragilis* (Pourtalès, 1868)	w	S	A	F	w:75-822
93		*Polymyces wellsi* Cairns, 1991	c	S	A	F	w:549-1682
94		*Pourtalocyathus hispidus* (Pourtalès, 1878)	w	S	A	-	w:349-1006
95		*Pourtalosmilia anthophyllites* (E. & S., 1786)	e	C	A	C	e:250-300
96		*Premocyathus cornuformis* (Pourtalès, 1868)	aac	S	F	C	w:137-1290; e:1300-2360
97		*Premocyathus dentiformis* (Alcock, 1902)	c	S	F	C	w:231-258
98		*Rhizosmilia gerdae* Cairns, 1978	w	C	A	C	w:123-549
99		*Rhizosmilia maculata* (Pourtalès, 1874)	w	C	A	C	w:0.5-508
100		*Schizocyathus fissilis* Pourtalès, 1874	aad	S	F	-	w:88-640; e:410-1300
101		*Solenosmilia variabilis* Duncan, 1873	c	C	A	C	w:220-1383; e:860-2165
102		*Stenocyathus vermiformis* (Pourtalès, 1868)	c	S	F	-	w:165-835; e:110-1229
103		*Stephanocyathus coronatus* (Pourtalès, 1867)	c	S	F	C	w:543-1250
104		*Stephanocyathus crassus* (Jourdan, 1895)	e	S	F	C	e:400-1557
105		*Stephanocyathus diadema* Moseley, 1876	w	S	F	C	w:795-2553
106		*Stephanocyathus laevifundus* Cairns, 1977	w	S	F	C	w:300-1158
107		*Stephanocyathus moseleyanus* (Sclater, 1886)	e	S	F	C	e:1000-2000
108		*Stephanocyathus nobilis* (Moseley, 1873)	e	S	F	C	e:1300-2200
109		*Stephanocyathus paliferus* Cairns, 1977	w	S	F	C	w:220-715
110		*Tethocyathus cylindraceus* (Pourtalès, 1868)	w	S	A	C	w:183-649
111		*Tethocyathus recurvatus* (Pourtalès, 1878)	w	S	F	C	w:320-569
112		*Tethocyathus variabilis* Cairns, 1979	aad	S	A	C	w:250-576; e:269-860
	FLA	*Thalamophyllia gombergi* Cairns, 1979	w	C	A	C	w:188-220
113		*Thalamophyllia riisei* (Duch. & Mich., 1860)	w	C	A	C	w:4-914
114		*Thecopsammia socialis* Pourtalès, 1868	w	S	A	D	w:214-878
	CUB	*Trematotrochus corbicula* (Pourtalès, 1878)	w	S	F	-	w:400-576
	NWC	*Trochocyathus fasciatus* Cairns, 1979	w	S	A	C	w:238
115		*Trochocyathus fossulus* Cairns, 1979	w	S	A	C	w:205-775
	BZL	*Trochocyathus laboreli* Cairns, 2000	w	S	A	C	w:130-240
116		*Trochocyathus rawsonii* Pourtalès, 1874	w	S	A	C	w:55-700
117		*Trochocyathus spinosocostatus* Zibrowius, 1980	e	S	A	C	e:1575-1650
118		*Trochopsammia infundibulum* Pourtalès, 1878	w	S	A	D	w:532-1472
119		*Vaughanella concinna* Gravier, 1915	c	S	A	C	e:1022-3018
	CTC	*Vaughanella margaritata* (Jourdan, 1895)	w	S	A	C	w:1267

Table 2: Description and abbreviations of the 38 North Atlantic regions used in cluster analysis, ordered clockwise from west to east. Number of deep-water (over 200 m) species listed in parentheses for each region. Two outlier regions omitted from ordination are bracketed.

[SPP]	St. Peter and Paul Rocks (6 species)
BZL	Tropical coast of Brazil, including coast south of equator to Rio de Janeiro (32)
GUY	Guyana, Suriname, French Guiana (22)
SC	Southern Caribbean: coasts of Venezuela and Colombia (29)
SWC	Southwestern Caribbean: offshore islands and banks from Panama to Nicaragua (30)
NWC	Northwestern Caribbean: offshore islands and banks from Honduras to Yucatan Channel (45)
WGM	Western Gulf of Mexico (36)
EGM	Eastern Gulf of Mexico (45)
LAV	Lesser Antilles (off Venezuela): Los Testigos to Aruba (27)
LA	Lesser Antilles: Virgin Islands to Trinidad (71)
HPR	Hispaniola and Puerto Rico (35)
CUB	Cuba and Cayman Islands (70)
JAM	Jamaica (43)
BAH	Bahamas (64)
FLA	Eastern Florida and Florida Keys (53)
GSN	Southeastern U.S: Georgia, S. Carolina, N. Carolina to Cape Hatteras (40)
BER	Bermuda and Muir Seamount (19)
CTU	Cold temperate U.S. coast from Cape Hatteras to Gulf of Maine, including Georges Bank (15)
CTC	Cold temperate Canadian coast from Bay of Fundy to Cape Breton Island, including: Browns Bank, Sable Bank, Grand Bank and Orphan Knoll (10)
[ARC]	Arctic: off Labrador, Newfoundland, Greenland (3)
ICE	Iceland, Reykjanes Ridge (9)
SCA	Scandinavia: coast of Norway (6)
SHE	Shetland, Faroe, Orkney, New Hebrides Islands; Rockall Bank (19)
GB	North Sea and coasts of Great Britain, Ireland and France (Bay of Biscay) to 45°N (29)
SPA	Southern Bay of Biscay (northern coast of Spain) (34)
POR	West coast of Spain and Portugal (43°-37°N) (34)
IMG	Ibero-Moroccan Gulf (37°-33°N) (33)
MED	Mediterranean (21)
MWS	Morocco and Western Sahara (33°-22°N) (31)
AZO	Azores (51)
MAD	Madeira (41)
CAN	Canary Islands (15)
CVI	Cape Verde Islands (14)
SMA	Seamounts and banks south of Azores: Pluto, Atlantis, Cruiser, Irving, Hyeres, Great Meteor (17)
SMP	Seamounts and banks between Portugal and Madeira: Gorringe, Gettysburg, Ampere, Seine, Josephine (22)
MAU	Coasts of Mauritania and Senegal (22°-13°N) (15)
GG	Gulf of Guinea (18)
IP	Indo-Pacific: species having distributions outside Atlantic Ocean (23)

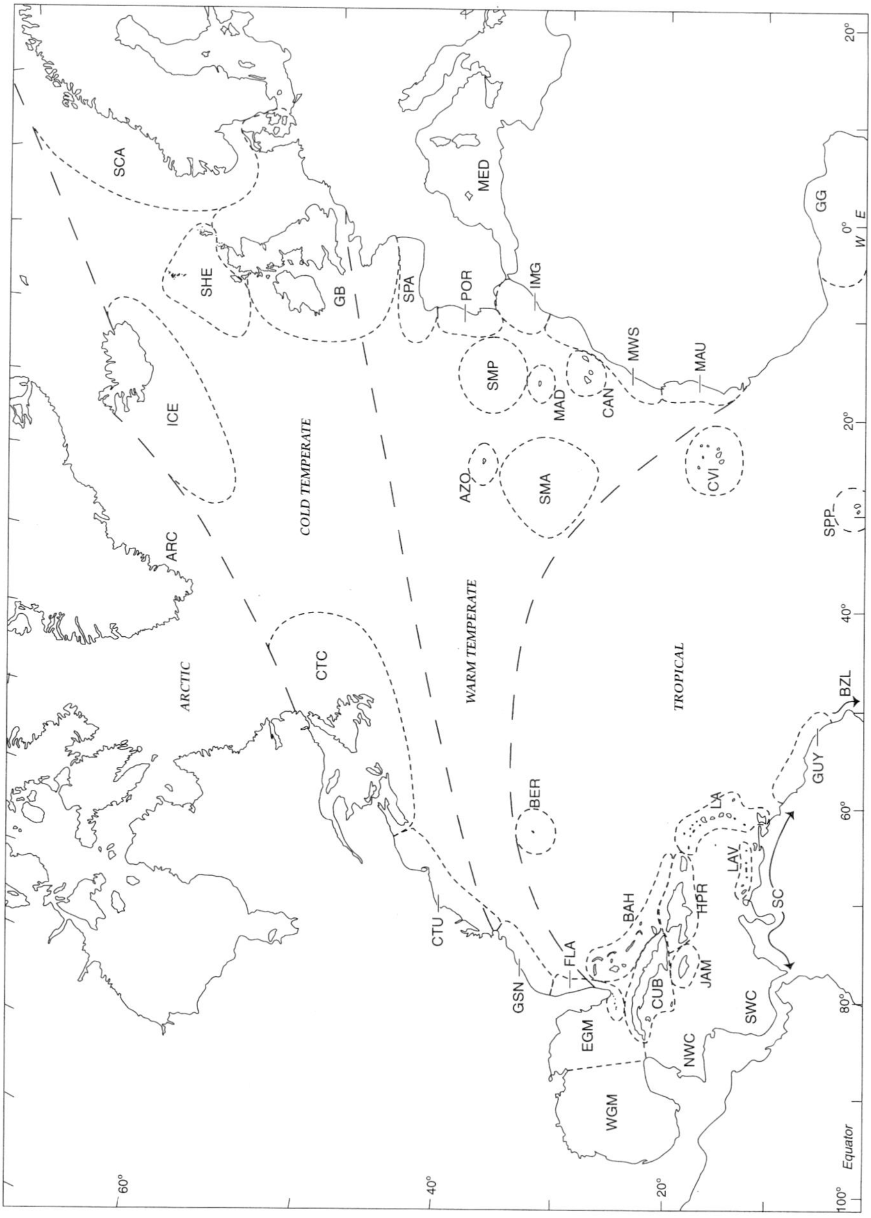

Figure 1. Locator map of North Atlantic showing 37 of the 38 regions scored in the analysis. Indo-Pacific region not mapped.

Table 3. The data matrix. The 38 rows correspond to the localities listed in Table 2. The 119 columns correspond to the species listed in Table 1.

Locale/Taxa	1	2	3	4	5	6	7	8	9	10	11	12	13	14	15	16	17	18	19	20	21	22	23	24	25	26	27	28	29	30
1 BER	0	0	0	0	0	0	0	0	0	0	0	0	0	0	0	0	0	1	0	0	0	0	0	0	0	0	1	0	0	0
2 GSN	0	0	1	0	1	0	0	0	0	1	0	0	0	1	1	0	0	1	0	0	0	0	0	0	0	0	0	0	0	0
3 FLA	0	0	1	1	1	0	0	0	0	1	0	0	0	1	1	0	0	1	0	0	0	1	0	0	0	1	0	0	0	0
4 BAH	1	1	1	1	0	0	0	1	0	0	1	0	1	0	1	0	0	1	1	0	0	1	0	1	0	0	0	0	0	0
5 WGM	0	0	1	0	1	0	0	0	0	0	1	0	0	0	1	0	0	1	0	0	1	1	0	0	0	1	0	0	0	0
6 EGM	1	0	1	0	1	0	0	0	0	1	0	0	0	0	0	0	0	1	0	0	0	1	0	0	0	0	0	0	0	0
7 CUB	1	0	0	1	0	0	0	1	0	1	1	0	1	1	0	0	0	1	1	0	0	1	0	1	0	1	0	0	0	1
8 HPR	1	0	0	0	0	0	0	0	0	0	0	0	0	0	0	0	0	1	1	0	0	1	0	0	0	0	0	0	0	0
9 JAM	1	0	0	0	0	0	0	0	0	0	1	0	1	1	0	0	0	1	0	0	0	1	0	0	0	1	0	0	0	1
10 LA	1	0	1	1	0	0	0	1	1	1	1	0	1	0	0	0	0	1	1	0	1	1	0	1	0	1	0	0	0	1
11 LAV	1	1	1	0	0	0	0	0	1	0	0	0	0	0	0	0	0	1	0	0	0	1	0	0	0	0	0	0	0	0
12 SC	1	1	1	0	0	0	0	0	1	0	0	0	0	0	0	0	0	1	0	0	0	0	0	0	0	0	0	0	0	0
13 SWC	1	0	0	0	0	0	0	0	0	1	0	0	0	0	0	0	0	1	0	0	0	1	0	0	0	0	0	0	0	0
14 NWC	1	0	0	0	0	0	0	1	0	0	1	0	0	0	0	0	0	1	0	0	0	1	0	0	0	0	0	0	0	0
15 GUY	0	0	1	0	0	0	0	0	1	0	0	0	0	0	0	0	0	1	0	0	0	1	0	0	0	0	0	0	0	0
16 BZL	1	0	0	0	0	0	0	0	1	0	0	0	0	0	0	0	0	1	0	0	0	0	0	0	0	0	0	0	0	0
17 SPP	1	0	0	0	0	0	0	0	0	0	0	0	0	0	0	0	1	0	0	0	0	0	0	0	0	0	0	0	0	0
18 CTU	0	0	0	0	1	0	0	0	0	0	0	0	0	0	0	0	1	1	0	0	0	0	0	0	0	0	0	0	0	0
19 CTC	0	0	0	0	0	0	0	0	0	0	0	0	0	0	0	0	1	0	0	0	0	0	0	0	0	0	0	0	0	0
20 ARC	0	0	0	0	0	0	0	0	0	0	0	0	0	0	0	0	0	0	0	0	0	0	0	0	0	0	0	0	0	0
21 SCA	0	0	0	0	0	0	0	0	0	0	0	0	0	0	0	0	0	0	0	0	0	0	0	0	0	0	0	0	1	0
22 ICE	0	0	0	0	0	0	0	0	0	0	0	0	0	0	0	0	0	0	0	0	0	0	0	0	0	0	0	0	0	0
23 SHE	0	0	0	0	0	0	0	0	0	0	0	0	0	0	0	1	1	0	0	1	0	0	0	0	0	0	0	1	1	0
24 GB	0	0	0	0	0	0	1	0	0	0	0	0	0	0	0	1	1	0	0	1	0	0	1	0	0	0	1	1	1	0
25 SPA	1	0	0	0	0	1	1	0	0	0	0	1	0	0	0	1	1	0	0	1	0	0	1	0	1	0	1	1	1	0
26 POR	1	0	0	0	0	1	1	0	0	0	0	1	0	0	0	1	1	0	0	1	0	0	1	0	1	0	1	1	1	0
27 MED	0	0	0	0	0	0	1	0	0	0	0	0	0	0	0	0	1	0	0	1	0	0	1	0	1	0	1	0	1	0
28 IMG	0	0	0	0	0	1	1	0	0	0	0	1	0	0	0	1	1	0	0	0	0	0	1	0	1	0	0	1	1	0
29 MWS	0	1	1	0	0	0	0	0	0	0	0	0	0	0	0	1	1	0	0	1	0	0	1	0	0	0	0	1	1	0
30 MAU	0	1	0	0	0	0	0	0	0	1	0	0	0	0	0	0	0	0	0	0	0	0	0	0	0	0	0	1	1	0
31 GG	0	1	0	0	0	0	0	0	0	1	0	0	0	0	0	0	1	0	0	0	0	0	0	0	0	0	0	0	1	0
32 AZO	1	0	0	0	0	1	1	0	0	0	0	0	0	0	0	1	1	0	0	1	0	0	1	0	1	0	1	0	1	0
33 SMA	1	0	0	0	0	0	1	0	0	0	0	0	0	0	0	0	0	0	0	1	0	0	1	0	1	0	0	0	0	0
34 SMP	1	0	0	0	0	1	1	0	0	0	0	0	0	0	0	1	0	0	0	0	0	0	0	0	1	0	0	0	1	0
35 MAD	1	0	1	0	0	1	0	0	0	0	0	1	0	0	0	1	1	0	0	1	0	0	1	0	1	0	1	0	0	0
36 CAN	1	0	1	0	0	0	0	0	0	0	0	0	0	0	0	0	0	0	0	0	0	0	0	0	1	0	0	1	1	0
37 CVI	0	0	0	0	0	0	0	0	0	0	0	0	0	0	0	0	0	0	0	0	0	0	0	0	1	0	0	1	0	0
38 IP	0	1	0	0	0	0	0	0	0	0	0	0	0	0	0	0	1	0	0	0	0	0	0	0	0	0	0	0	0	0

Table 3. The data matrix *(continued)*

Locale/Taxa	31	32	33	34	35	36	37	38	39	40	41	42	43	44	45	46	47	48	49	50	51	52	53	54	55	56	57	58	59	60
1 BER	0	0	0	0	0	0	0	1	0	0	0	0	0	0	1	1	1	1	0	0	0	1	0	0	0	0	1	0	0	0
2 GSN	0	1	0	0	0	0	0	0	1	1	1	0	1	0	1	1	1	0	1	0	0	1	0	0	1	1	1	1	1	0
3 FLA	0	1	1	0	0	0	0	0	1	1	1	1	1	0	1	1	1	1	1	0	0	1	0	1	0	1	0	0	0	0
4 BAH	0	0	1	0	0	1	1	1	0	0	0	0	0	1	1	1	1	1	1	1	0	1	1	1	1	1	1	0	0	1
5 WGM	0	0	1	0	0	1	1	0	0	0	0	0	1	0	1	1	1	0	0	1	0	0	0	0	1	0	0	0	0	0
6 EGM	0	1	1	0	0	1	1	0	1	0	1	1	1	0	1	1	1	0	0	0	0	0	0	1	1	1	0	0	0	0
7 CUB	0	0	1	0	0	1	1	1	0	0	0	0	1	1	1	1	1	1	1	1	0	1	1	1	1	1	1	0	0	1
8 HPR	0	0	0	0	0	0	1	1	0	0	0	0	0	1	1	1	1	0	0	0	0	0	1	0	0	0	1	0	0	0
9 JAM	0	0	0	0	0	1	0	1	0	0	0	0	0	0	1	1	1	0	0	1	0	0	1	0	1	0	1	0	0	0
10 LA	0	0	0	0	0	1	1	1	0	0	0	0	1	1	1	1	1	1	0	1	0	1	1	1	1	1	1	0	0	0
11 LAV	0	0	0	0	0	0	1	1	0	0	1	1	0	0	1	0	1	0	0	0	0	0	0	1	0	0	0	0	0	0
12 SC	0	1	0	0	0	1	1	0	0	0	0	0	0	0	1	1	1	0	0	1	0	0	0	1	1	0	0	0	0	0
13 SWC	0	1	0	0	0	0	1	1	0	0	0	0	0	0	1	1	0	0	0	0	0	0	0	0	0	0	1	0	0	0
14 NWC	0	1	1	0	0	1	1	1	1	0	0	0	0	0	1	1	1	1	0	0	0	0	0	1	1	0	0	0	0	0
15 GUY	0	0	0	0	0	0	1	0	0	0	1	0	0	0	1	0	0	0	0	0	0	0	0	0	1	0	0	0	0	0
16 BZL	0	1	1	0	0	1	0	0	0	0	0	1	1	0	1	1	1	0	0	0	0	1	0	0	0	0	1	0	0	0
17 SPP	0	1	0	0	0	0	0	0	0	0	0	0	0	0	0	0	0	0	0	0	0	0	0	0	1	0	0	0	0	0
18 CTU	0	0	0	0	0	0	0	0	0	0	1	0	0	0	0	0	1	0	0	0	0	1	0	0	0	1	1	1	1	0
19 CTC	0	0	0	0	0	0	0	0	0	0	1	0	0	0	0	0	0	0	0	0	0	1	0	0	0	0	0	1	1	0
20 ARC	0	0	0	0	0	0	0	0	0	0	0	0	0	0	0	0	0	0	0	0	0	0	0	0	0	0	0	1	0	0
21 SCA	0	0	0	0	0	0	0	0	0	0	0	0	0	0	0	0	0	0	0	0	0	1	0	0	0	0	0	0	0	0
22 ICE	0	0	0	0	0	0	0	0	0	0	0	0	0	0	0	0	0	0	0	0	0	1	0	0	0	0	0	0	0	0
23 SHE	0	0	0	0	0	0	0	0	0	0	0	0	0	0	0	0	0	0	0	0	0	1	0	0	0	0	0	1	1	0
24 GB	0	0	0	0	0	0	0	0	0	0	0	0	0	0	0	0	0	1	0	0	1	1	0	1	0	0	1	1	1	0
25 SPA	0	0	0	0	0	0	0	0	0	0	0	0	0	0	0	0	1	1	0	1	1	1	0	1	0	0	1	1	1	0
26 POR	0	0	0	1	0	0	0	0	0	0	1	0	0	0	0	0	0	1	0	0	1	1	0	0	0	0	1	1	1	0
27 MED	1	1	0	1	1	0	0	0	0	0	0	0	0	0	0	0	0	0	0	0	1	1	0	0	0	0	0	0	0	0
28 IMG	0	1	0	0	1	0	0	0	0	0	1	0	0	0	0	0	1	0	0	0	1	1	0	1	1	0	0	1	1	0
29 MWS	0	1	0	1	1	0	0	0	0	0	1	0	0	0	0	1	1	0	0	0	1	1	0	0	0	0	1	0	1	0
30 MAU	1	1	1	1	1	0	0	0	0	0	0	0	0	0	0	0	0	0	0	0	1	0	0	0	0	0	0	0	0	0
31 GG	0	1	1	0	1	0	0	0	0	0	0	0	0	0	0	0	1	0	0	0	0	0	0	0	1	0	1	1	0	0
32 AZO	0	0	0	0	0	0	0	0	1	0	1	1	1	0	0	1	1	1	0	1	1	1	0	0	0	0	1	1	1	0
33 SMA	0	0	0	0	0	0	0	0	0	0	1	0	1	0	0	1	0	1	0	0	0	0	0	0	1	0	1	0	0	0
34 SMP	0	0	0	0	0	0	0	0	0	0	0	0	1	0	0	1	0	1	0	0	1	1	0	0	0	0	0	1	0	0
35 MAD	0	1	0	0	0	0	0	0	0	0	1	0	1	0	0	1	1	1	0	0	0	1	0	0	1	0	1	1	1	0
36 CAN	0	1	0	0	0	0	0	0	0	0	0	0	0	0	0	0	0	0	0	0	1	1	0	0	0	0	0	1	0	0
37 CVI	0	1	1	0	0	0	0	0	0	0	0	1	0	0	0	1	0	0	0	0	1	0	0	0	0	0	0	1	0	0
38 IP	0	0	0	0	0	0	0	0	0	0	1	1	0	0	0	0	0	0	0	0	0	1	0	0	1	0	1	0	0	0

Table 3. The data matrix *(continued)*

Locale/Taxa	61	62	63	64	65	66	67	68	69	70	71	72	73	74	75	76	77	78	79	80	81	82	83	84	85	86	87	88	89	90
1 BER	0	0	0	0	0	0	0	0	1	0	0	1	1	0	0	0	0	1	0	1	0	0	1	0	1	0	1	0	0	0
2 GSN	0	0	1	0	0	0	0	0	0	0	0	0	1	1	1	1	0	1	0	1	0	0	1	1	1	0	1	0	0	0
3 FLA	0	1	0	1	1	0	0	1	1	1	0	0	1	0	1	1	0	1	0	1	0	0	1	1	0	0	1	1	0	0
4 BAH	0	1	0	0	0	0	1	1	1	0	1	1	1	1	0	0	0	1	0	1	1	0	1	1	1	0	1	0	1	0
5 WGM	0	1	0	0	0	0	0	0	0	0	0	1	1	0	0	0	0	1	1	1	1	0	1	1	1	0	1	0	0	0
6 EGM	0	1	0	1	1	0	0	0	0	0	0	1	1	0	0	0	0	1	1	1	1	0	1	1	1	0	1	0	1	0
7 CUB	0	0	0	1	0	0	1	0	1	0	1	1	1	0	0	1	1	0	1	1	1	0	1	1	1	0	1	1	1	0
8 HPR	0	0	0	0	0	0	1	0	1	0	1	1	1	0	0	0	0	1	1	1	0	0	1	1	1	0	1	0	1	0
9 JAM	0	0	0	1	0	0	0	0	1	1	1	1	1	1	0	0	0	1	0	1	1	0	1	1	0	0	1	0	1	0
10 LA	0	0	0	1	1	0	0	1	1	0	1	1	1	0	0	1	0	1	1	1	1	0	1	1	1	0	1	1	1	1
11 LAV	0	0	0	1	0	0	0	0	0	0	0	1	0	0	0	0	0	0	1	1	0	0	0	1	1	0	1	0	0	0
12 SC	0	0	0	1	0	0	0	0	0	0	0	0	1	0	0	0	0	1	1	1	1	0	0	1	1	0	1	0	0	0
13 SWC	0	0	0	1	0	0	0	0	0	0	0	1	1	0	0	0	0	0	1	1	1	0	1	0	1	0	1	0	1	0
14 NWC	0	0	0	1	1	0	0	1	0	1	1	1	1	0	0	1	0	0	1	1	1	0	1	1	0	0	1	0	1	0
15 GUY	0	0	0	0	0	0	0	0	0	0	0	1	1	0	0	0	0	0	1	1	0	0	0	1	1	0	1	0	1	0
16 BZL	0	0	0	0	1	0	0	0	1	0	0	0	1	0	0	0	0	1	1	0	1	0	0	1	0	0	1	0	1	1
17 SPP	0	0	0	0	0	0	0	0	0	0	0	0	0	0	0	0	0	0	0	0	0	0	1	0	0	0	0	0	0	0
18 CTU	0	0	1	0	0	1	0	0	0	0	0	0	1	0	0	0	0	1	0	0	0	0	0	0	0	0	0	0	0	0
19 CTC	0	0	1	0	0	0	0	0	0	0	0	0	1	0	0	0	0	1	0	0	0	0	0	0	0	0	0	0	0	0
20 ARC	0	0	0	0	0	0	1	0	0	0	0	0	0	0	0	0	0	0	0	0	0	0	0	0	0	0	0	0	0	0
21 SCA	0	0	1	0	0	1	0	0	0	0	0	0	0	0	0	0	0	1	0	0	0	0	0	1	0	0	0	0	0	0
22 ICE	0	0	0	0	0	1	1	0	0	0	0	0	0	0	0	0	0	1	0	0	0	0	0	1	0	0	0	0	0	0
23 SHE	0	0	1	0	0	1	1	0	0	0	0	0	0	0	0	0	0	1	0	0	0	0	0	1	0	0	0	0	0	0
24 GB	0	0	1	0	0	1	1	0	0	0	0	0	1	0	0	0	0	1	0	0	0	0	0	1	0	0	0	0	0	0
25 SPA	0	0	1	0	0	0	1	0	0	0	0	0	1	0	0	0	0	1	0	0	0	0	0	1	0	0	0	0	0	0
26 POR	1	0	0	0	1	1	1	0	0	0	0	0	0	0	0	0	0	1	0	0	0	0	0	1	0	0	1	0	0	0
27 MED	0	0	0	0	0	0	0	0	0	0	0	1	1	0	0	0	0	1	0	0	1	0	0	1	0	0	1	0	0	0
28 IMG	1	0	1	0	0	1	0	0	0	0	0	0	1	0	0	0	0	1	0	0	0	0	0	1	0	0	1	0	0	1
29 MWS	1	0	0	0	0	1	1	0	0	0	0	0	1	0	0	0	0	1	0	0	0	0	0	1	0	0	1	0	0	0
30 MAU	0	0	1	0	0	0	0	0	0	0	0	0	0	0	0	0	0	1	0	0	0	0	0	1	0	0	0	0	0	0
31 GG	0	0	0	0	0	0	1	0	0	0	0	0	0	0	0	0	0	1	0	0	0	0	0	1	0	0	0	0	0	0
32 AZO	1	0	1	0	1	1	1	0	0	0	0	1	1	1	0	0	0	1	0	0	1	1	0	1	0	0	1	1	0	0
33 SMA	1	0	0	0	0	0	0	0	0	0	0	0	0	0	0	0	0	0	0	0	0	1	0	1	0	1	1	0	0	0
34 SMP	1	0	0	0	1	0	0	0	0	0	0	0	0	0	0	0	0	1	0	0	0	0	0	1	0	1	1	1	0	0
35 MAD	1	0	0	0	1	1	0	0	0	0	0	1	1	0	0	0	0	1	1	0	1	1	0	1	0	0	1	1	0	0
36 CAN	0	0	0	0	0	0	0	0	0	0	0	0	0	0	0	0	0	1	1	0	1	0	0	0	0	0	1	0	0	0
37 CVI	0	0	0	0	0	1	1	0	0	0	0	0	0	0	0	0	0	1	1	0	0	0	0	1	0	0	1	0	0	0
38 IP	0	0	0	0	0	1	1	0	0	1	0	1	1	0	0	0	1	1	0	0	1	0	0	1	0	0	0	0	0	0

Table 3. The data matrix *(continued)*

Locale/Taxa	91	92	93	94	95	96	97	98	99	100	101	102	103	104	105	106	107	108	109	110	111	112	113	114	115	116	117	118	119
1 BER	0	0	0	0	0	1	0	0	0	0	1	0	0	0	0	0	0	0	0	0	0	0	0	0	0	0	0	0	0
2 GSN	0	1	0	0	0	1	0	0	0	0	1	1	0	0	1	0	0	0	0	0	0	0	0	1	0	1	0	0	0
3 FLA	0	1	0	0	0	1	0	0	1	1	1	1	1	0	1	1	0	0	1	0	0	1	0	1	0	1	0	0	0
4 BAH	0	1	1	1	0	0	0	1	1	1	0	1	1	0	1	1	0	0	1	1	1	1	1	1	1	1	0	0	0
5 WGM	1	1	0	0	0	0	0	0	0	1	0	1	0	0	0	0	0	0	1	1	0	0	1	0	0	1	0	0	0
6 EGM	0	1	0	0	0	1	0	0	0	1	1	1	1	0	1	0	0	0	1	0	0	0	1	0	0	1	0	1	0
7 CUB	1	1	1	1	0	1	0	1	1	1	1	1	1	0	1	1	0	0	1	0	1	1	1	0	1	1	0	1	0
8 HPR	1	0	0	1	0	0	0	1	1	1	0	0	1	0	1	1	0	0	0	0	0	1	0	0	0	1	0	0	0
9 JAM	1	0	0	1	0	1	0	0	1	0	1	0	1	0	1	0	0	0	1	1	1	0	1	0	0	1	0	0	0
10 LA	1	1	0	1	0	1	1	1	1	1	1	1	1	0	1	1	0	0	1	1	0	1	1	0	1	1	0	1	0
11 LAV	0	0	0	0	0	1	0	0	1	1	1	0	1	0	1	0	0	0	1	0	0	0	0	0	0	0	0	0	0
12 SC	0	1	0	0	0	0	0	0	1	0	0	0	1	0	1	0	0	0	1	0	0	0	0	0	0	1	0	0	0
13 SWC	0	1	1	0	0	0	0	0	1	0	0	0	1	0	1	1	0	0	1	0	0	0	1	0	0	1	0	0	0
14 NWC	1	1	0	0	0	1	0	0	1	0	0	0	1	0	1	0	0	0	1	0	0	1	1	0	0	1	0	0	0
15 GUY	0	0	0	0	0	0	0	0	0	1	1	0	0	0	1	0	0	0	1	0	0	0	1	0	0	1	0	0	0
16 BZL	0	1	0	0	0	1	0	0	1	0	1	1	0	0	1	0	0	0	1	0	0	0	0	0	0	1	0	0	0
17 SPP	0	0	0	0	0	0	0	0	0	0	0	1	0	0	0	0	0	0	0	0	0	0	0	0	0	0	0	0	0
18 CTU	0	0	0	0	0	0	0	0	0	0	1	0	0	0	0	0	0	0	0	0	0	0	0	0	0	0	0	0	0
19 CTC	0	0	0	0	0	1	0	0	0	0	0	0	0	0	0	0	0	0	0	0	0	0	0	0	0	0	0	0	0
20 ARC	0	0	0	0	0	0	0	0	0	0	1	0	0	0	0	0	0	0	0	0	0	0	0	0	0	0	0	0	0
21 SCA	0	0	0	0	0	0	0	0	0	0	0	0	0	0	0	0	0	0	0	0	0	0	0	0	0	0	0	0	0
22 ICE	0	0	0	0	0	1	0	0	0	0	1	0	0	1	0	0	1	0	0	0	0	0	0	0	0	0	0	0	0
23 SHE	0	0	0	0	0	0	0	0	0	0	1	1	0	0	0	0	1	1	0	0	0	0	0	0	0	0	0	0	1
24 GB	0	0	0	0	0	1	0	0	0	0	1	1	0	0	0	0	1	1	0	0	0	0	0	0	0	0	0	0	1
25 SPA	0	0	0	0	0	1	0	0	0	0	1	1	0	1	0	0	1	1	0	0	0	0	0	0	0	0	0	0	1
26 POR	0	0	0	0	1	1	0	0	0	1	1	1	0	0	0	0	1	1	0	0	0	0	0	0	0	0	0	0	0
27 MED	0	0	0	0	1	0	0	0	0	0	0	1	0	0	0	0	0	0	0	0	0	0	0	0	0	0	0	0	0
28 IMG	0	0	0	0	0	1	0	0	0	1	1	1	0	0	0	0	1	1	0	0	0	0	0	0	0	0	0	0	0
29 MWS	0	0	0	0	1	1	0	0	0	0	1	0	0	0	0	0	1	1	0	0	0	1	0	0	0	0	0	0	0
30 MAU	0	0	0	0	1	0	0	0	0	0	0	0	0	0	0	0	0	0	0	0	0	0	0	0	0	0	0	0	0
31 GG	0	0	0	0	1	0	0	0	0	0	1	0	0	0	0	0	0	1	0	0	0	0	0	0	0	0	1	0	0
32 AZO	0	0	0	0	1	1	0	0	0	1	1	1	0	1	0	0	1	1	0	0	0	1	0	0	0	0	1	0	1
33 SMA	0	0	0	0	0	0	0	0	0	0	0	1	0	0	0	0	0	0	0	0	0	0	0	0	0	0	0	0	0
34 SMP	0	0	0	0	0	0	0	0	0	0	1	1	0	0	0	0	0	1	0	0	0	0	0	0	0	0	0	0	0
35 MAD	0	0	0	0	0	1	0	0	0	1	1	1	0	1	0	0	1	1	0	0	0	0	0	0	0	0	0	0	1
36 CAN	0	0	0	0	1	0	0	0	0	0	0	0	0	0	0	0	1	0	0	0	0	0	0	0	0	0	0	0	0
37 CVI	0	0	0	0	0	0	0	0	0	0	0	0	0	0	0	0	0	0	0	0	0	0	0	0	0	0	0	0	0
38 IP	0	0	1	0	0	0	1	0	0	0	1	1	1	0	0	0	0	0	0	0	0	0	0	0	0	0	0	0	1

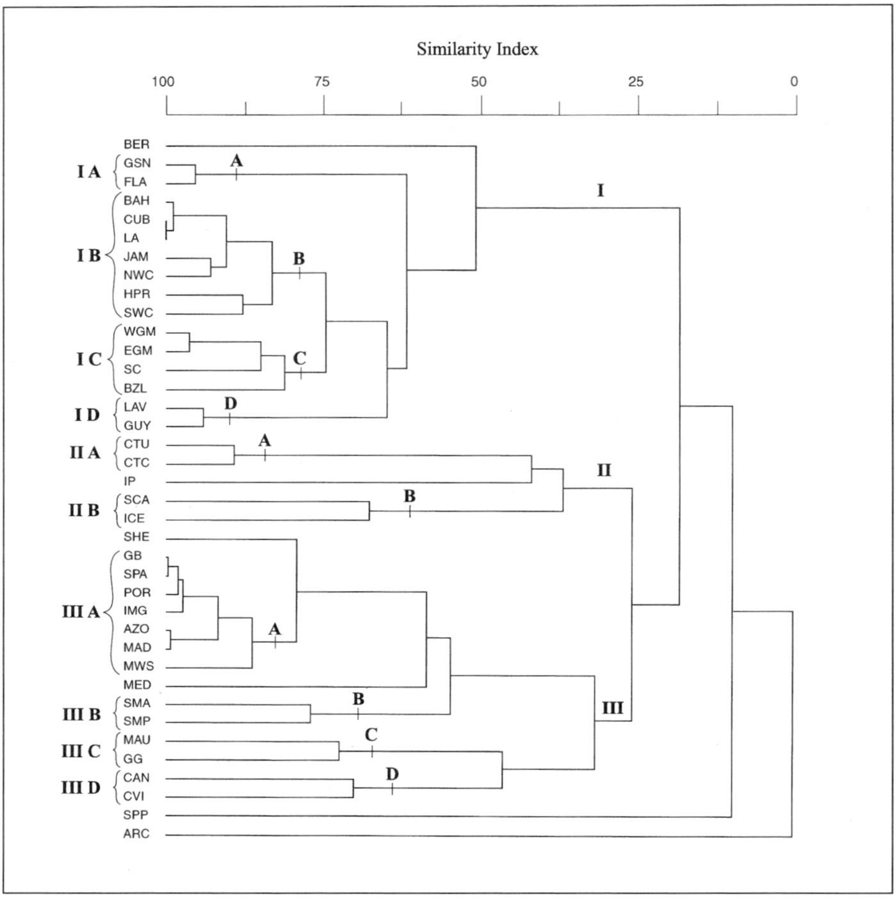

Figure 2. The dendrogram of the 38 regions produced by the UPGMA clustering, the superclusters indicated by Roman numerals, the component clusters by upper case letters. Percent chaining = 19.9. Percent similarity indicated on scale at top.

the progressively less useful values to the right of the dendrogram that result from averaging together values for more and more localities, the arbitrary fashion in which dissimilar regions may be placed adjacent to one another (the brackets can rotate freely at each junction), and the fact that a cluster analysis will generate a cluster regardless of the strength of the connection (it is, as always, up to the researcher to recognize the usefulness of the result). In order to better visualize the relationships between the groups identified using the dendrograms and to seek an independent confirmation of those results (an indication of the robustness of those patterns), it is common, and recommended, to perform an ordination on the same data and using the same distance matrix.

An ordination presents the data in the original distance matrix, here 36 dimensional data, in a very reduced number of dimensions, typically two or three, depending on the patterns of variation present. Although this tends to introduce some distortion from reducing the dimensionality of the data, and also tends to lose some of the information present in the distance matrix, most analyses will produce ordinations that account for most of the variance present in the data, and will also show the major trends that are present. This was the case for our analyses. The two types of analyses are also complementary in that the groupings indicated by the cluster analysis can be superimposed on the continuously distributed ordination results. If the clusters can be superimposed with relatively little effort, then the results are considered relatively robust. If the clusters cross each other and are very distorted, the results are given less weight.

Many different methods are available for ecological/biogeographic ordinations, including Principal Components Analysis (PCA), Correspondence Analysis (CA), and Detrended Correspondence Analysis (DCA) (Sneath and Sokal, 1973; Clarke and Warwick, 1994). The ordination algorithm chosen to analyze these data was Non-Metric Multi-Dimensional Scaling (MDS, sometimes NMS or NMMDS). Simply stated, MDS produces a map (usually of two or three dimensions) of the analyzed samples (i.e., regions) that best represents the dissimilarity among all samples. It does this in a non-metric way, in that it converts all the Bray-Curtis similarity values to rank values, not absolute values, and produces the plot on this basis. It starts out with a configuration of the data in space and progressively moves points around. At each step it measures the goodness of fit of the data with the structure of the starting distance matrix and stops when the fit is no longer improved with additional movement. According to Clarke and Warwick (1994) and McCune and Mefford (1999), MDS is undoubtedly the best ordination technique to use for ecological data, and it is becoming increasingly popular in biogeographic analyses, especially now that computing power has increased so greatly over the past decade and has caught up with the needs of researchers. One of the few disadvantages of using MDS in the past was the long computation time needed to apply it. However, the software used in these analyses for both clustering and ordination, PC-ORD4 (McCune and Mefford, 1999), handled the data expediently. The MDS analysis, run on the autopilot mode on the "slow and thorough" option, took approximately 11 minutes, which included 400 iterations and 90 runs through the data: 40 with real data and 50 with randomized data.

RESULTS

The total starting data matrix of 38 localities by 134 taxa (5092 cells) is presented in Table 3. The number of species-presence records (1's in the matrix) is 1127, or 22.1% of the total. Although this is thought to be a good approximation of the distribution of these species, this percentage will certainly rise with additional collection as will the number of taxa. Table 1 presents data for each species: its general pattern of distribution, depth range in the west-

Table 4. Some of the characteristics of the various clusters, unclustered regions and outlying regions, as determined by Bray-Curtis clustering and MDS. The first number of every pair is the absolute number of species in the cluster having this quality, the second number the percentage this number constitutes of the total number of species in this cluster. Numbers in brackets are total numbers for the cluster or region; numbers in parentheses are endemic species. Abbreviations as in Table 1; cos – cosmopolitan.

	Zoogeographic Pattern								CFD Ratio		
Superclusters/Clusters	w	e	aad	aac	cos	over 1000 m	Colonial	Attached	C	F	D
North Atlantic [134]	61/46	26/19	20/15	4/3	23/17	50/37	30/22	86/64	75/56	12/9	23/17
Super cluster I: Western											
Atlantic [104]	60(60)/58	-	20/19	4/4	20/19	18/17	21/20	62/60	54/52	10/10	18/17
Cluster ID [32]	15/47	-	6/19	1/3	10/31	3/6	10/31	15/47	22/69	2/6	3/9
Cluster IB ("insular") [92]	54(19)/59	-	19/21	1/1	18/20	18/20	20/22	55/60	47/51	7/8	18/20
Cluster IC ("cont.") [60]	29(1)/48	-	15/25	1/ 2	15/25	4/7	18/30	34/57	32/53	4/7	2/20
Cluster IA [60]	29(4)/48	-	15/25	4/7	12/20	7/12	13/22	29/48	29/48	7/12	10/17
Bermuda [19]	7/37	-	5/26	1/5	6/31	5/26	6/32	12/63	13/68	1/5	1/5
Supercluster II: North											
Atlantic [22]	4(1)/18	3/14	1/5	4/18	10/45	16/73	6/27	10/45	12/55	4/18	2/9
Cluster IIA [17]	4(1)/24	-	1/6	4/22	8/47	12/71	5/29	8/47	9/53	5/29	2/12
Indo-Pacific [22]	-	-	-	-	22/100	8/36	6/27	10/45	10/45	2/9	2/9
Cluster IIB [11]	-	3/27	0/0	2/18	6/55	5/45	7/27	5/45	7/64	1/9	0/0
Supercluster III: Eastern											
Atlantic [68]	-	26(26)/38	20/29	4/6	18/26	32/47	17/25	38/56	36/53	7/10	13/19
Shetland [19]	-	7(1)/37	0/0	3/16	9/47	11/58	3/16	8/42	10/53	3/16	2/11
Cluster IIIA [62]	-	22(4)/35	18/29	4/6	18/29	32/52	16/26	31/50	33/53	7/11	9/15
Mediterranean [21]	-	10/48	3/14	0/0	8/38	3/14	8/38	17/81	12/57	1/5	2/10
Cluster IIIB [28]	-	12(1)/43	7/25	1/4	8/29	8/29	8/29	15/54	15/53	2/7	5/18
Cluster IIID [21]	-	6/29	6/29	1/5	8/38	6/29	9/43	11/52	13/62	1/5	2/10
Cluster IIIC [24]	-	10(1)/42	4/17	2/8	8/33	9/38	11/46	14/58	14/58	2/8	6/25
Outliers											
[St. Peter & Paul Rocks] [6]	1/17	0/0	2/33	0/0	3/50	1/17	4/67	2/33	3/50	0/0	1/17
[Arctic] [3]	0/0	0/0	0/0	1/33	2/67	3/100	1/33	1/33	1/33	0/0	1/33

ern and eastern Atlantic, coloniality, attachment, and whether the species is a caryophylliid, a flabellid, or a dendrophylliid, the three major taxonomic groups that occur at these localities.

The results of the UPGMA clustering are illustrated in Figures 2-3. After segregating the two outlier regions of St. Peter and Paul Rocks and the Arctic, the dendrogram shows three superclusters: western Atlantic, eastern Atlantic, and cold temperate North Atlantic. The western Atlantic grouping is the most highly supported supercluster, containing four discrete clusters and one unclustered outlier: Bermuda. The North Atlantic supercluster contains two component clusters. The least well-supported supercluster, the eastern Atlantic, contains four clusters and two outliers: the Shetland Islands and the Mediterranean. Table 4 lists various biological parameters that characterize those clusters and superclusters, as well as the outlier regions. Each cluster can be compared to the North Atlantic average (Table 4), i.e., the 134 species of deep-water scleractinians known from the North Atlantic. For instance, among all the North Atlantic deep-water corals, 46% are endemic to the western Atlantic, 19% endemic to the eastern Atlantic, 15% are disjunct amphi-Atlantic, 3% are continuous amphi-Atlantic, and 17% are widespread outside the Atlantic, if not cosmopolitan. Furthermore, 37% of this general group occur at depths greater than 1000 m (the deepest record being at 4822 m), 22% are colonial, 64% are attached, and the ratio of caryophylliids:flabellids:dendrophylliids is 56:9:17.

The results of the MDS ordination, done using the program PC-ORD4 (McCune and Mefford, 1999), are illustrated in Figure 4, with the clusters from the UPGMA cluster analysis overlain on the plot. The outliers (St. Peter and Paul Rocks, and the Arctic) are not shown and were not used in this analysis as they would have had an overly large effect on the results; the outlier analysis already indicated they are very different from the other localities and each other and we have interpreted the results with this in mind. The plot was also rotated 190° from the PC-ORD4 output to optimize the graphic representation of the results, a cosmetic change that has no effect on the inter-relationships of the points but makes the results graphically more congruent with the geography. The resultant two-dimensional ordination plot clearly shows the three superclusters and component clusters in a more graphic way than that obtained from the dendrogram. This combined presentation of the results forms the basis of much of the discussion that follows. In general, however, one can immediately see certain relationships that bear further analysis. For example, the component regions of cluster ID are not contiguous; Bermuda is placed closest to cluster IA; the outlier Mediterranean is central to the eastern Atlantic region; the North Atlantic is closer to the eastern than western Atlantic; and the Shetland Islands, although clustering with the eastern Atlantic, are transitional between the North and eastern Atlantic superclusters. All of these results are discussed in further detail below, resulting in the slightly modified MDS plot of Figure 5.

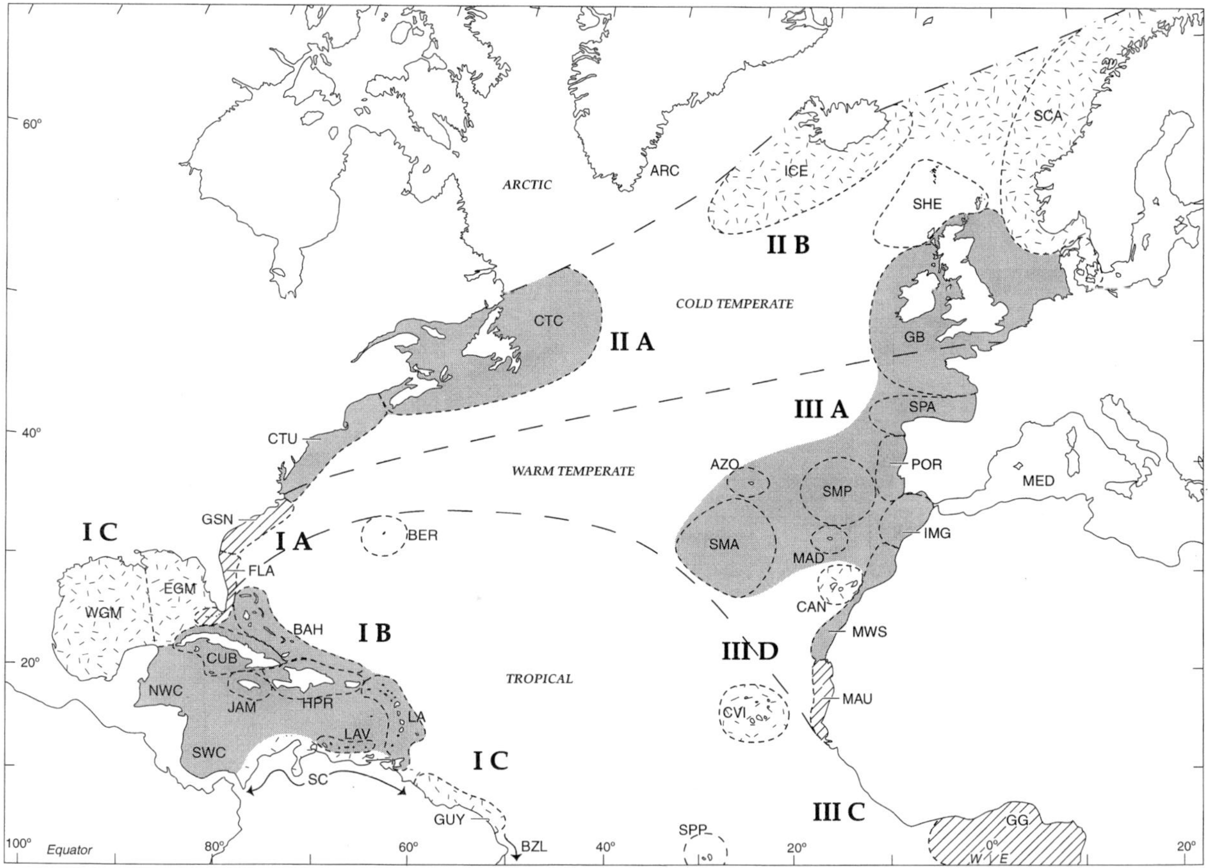

Figure 3. Map of the three major superclusters of North Atlantic deep-water Scleractinia and their component clusters, as explained in the text and Figure 5

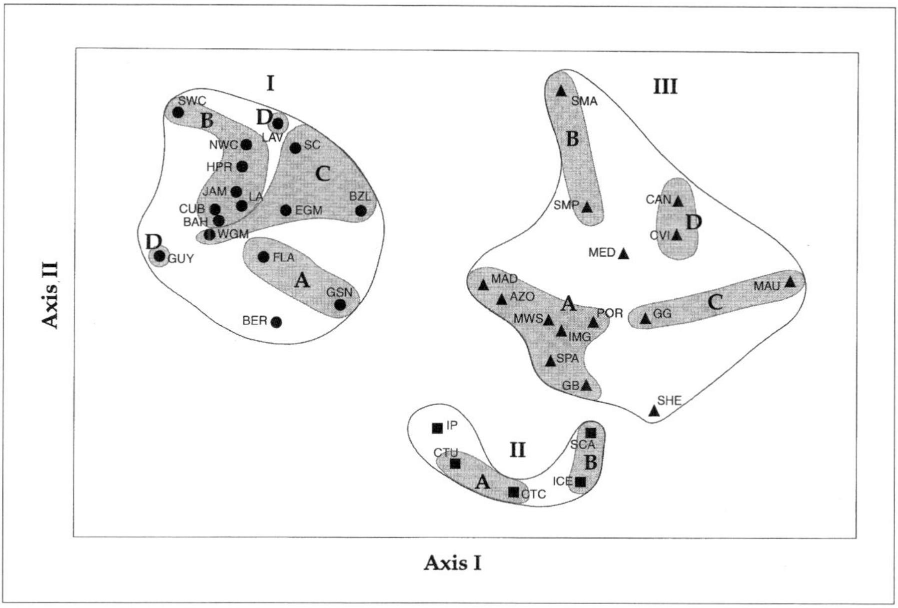

Figure 4. Multi-Dimensional Scaling plot of 36 regions, emphasizing the three superclusters and all component clusters of the UPGMA dendrogram (Figure 2)

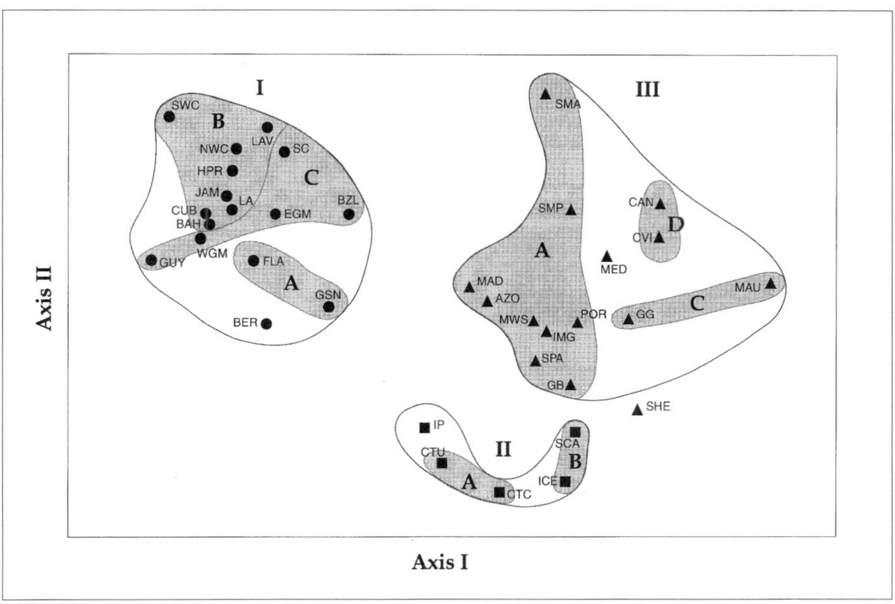

Figure 5. Multi-Dimensional Scaling plot of 36 regions, emphasizing three superclusters and a revised interpretation of the component clusters in superclusters I and III

DISCUSSION

Analysis of Clusters

The western Atlantic supercluster (I), consisting of 104 species that occur deeper than 200 m, is the best supported of the three major clusters in the analysis, containing 60 species (58%) endemic to the region (Table 4). Zoogeographically, it consists of the tropical and warm temperate western Atlantic realm (Briggs, 1974). Statistically it approximates the North Atlantic averages for most parameters, except for depth of capture; only 17% of its species occur deeper than 1000 m.

In an earlier analysis, Cairns (1979:205) distinguished several patterns of distribution of the deep-water corals of the western Atlantic, but stopped short of delimiting zoogeographic subregions characterized by endemic species, stating that the entire western Atlantic deep-water fauna was primarily one Caribbean province with depauperate extensions to the north and south. However, these analyses suggest that there are four deep-water subregions in the western Atlantic that can be defined and distinguished by species composition and various other parameters, as quantified in Table 4.

For instance, the largest cluster (IB), consisting of 92 species (including 19 species endemic to this cluster), unites the Greater and Lesser Antilles, as well as the far offshore islands and banks of the western Caribbean. Like the main cluster it belongs to, it is average in most parameters, including having a paucity of deep-water species, but geographically it consists of species that occur off islands. Even the records in the western Caribbean are found on the slopes of islands or banks often hundreds of km offshore (e.g., Arrowsmith Bank, Roatán, Rosalind Bank, Serrana Bank, Serranilla Bank). This insular pattern of distribution is similar to Ekman's (1953) Antillean region, Briggs' (1974) West Indian Province, Antillean pattern 2A of shallow-water azooxanthellates (Cairns, 2000), and deep-water coral distribution patterns 2 and 3 of Cairns (1979). It is also similar to the insular distribution pattern of stylasterids in the western Atlantic (Cairns, 1992), and may have a similar etiology, i.e., composed of species than require full oceanic salinity and clear water (not interrupted by low-salinity, sediment-laden water from riverine runoff) and an abundance of hard substrate (not muddy soft substrate).

A closely related cluster (IC), consisting of 60 species, groups the disjunct regions of the continental margins of Brazil, northern South America, and the Gulf of Mexico (Figures 2-3). It is essentially a subset of cluster IB, having only four species different from that cluster (*Astrangia poculata*, *Dasmosmilia lymani*, *D. variegata* and *Trochocyathus laboreli*, the last endemic to the subregion), and statistically (Table 4) is similar to cluster IB, except in having an even lower percentage of deep-living species. This pattern does not correlate well with traditional zoogeographic regions based on shallow-water organisms, for instance including three provinces described by Briggs (Brazilian, Caribbean and Carolinian), but does correspond to continental pattern 3B of shallow-water azooxanthellate corals (Cairns, 2000). It is suggested that the species of this "continental cluster" are more resistant to salinity and sediment fluctuations and perhaps more able to exist on soft substrates, conditions that are more common off continental margins than an insular setting.

Cluster ID groups two disjunct and disparate geographic areas (the islands off Venezuela and the continental slope off Guyana and Suriname), regions not only separated geographically but ecologically. Logically, the islands off Venezuela should group with the insular cluster, and in fact have a complete overlap in species except for two, and the Guyana/Suriname region should cluster with the continental cluster, and in fact all the species from that region do occur in the continental cluster. Thus, cluster ID is considered to be invalid, a mathematical anomaly of the clustering algorithm perhaps created by the relative low number of species in these regions and perhaps reflecting inadequate sampling of those regions. Based on their geography and species composition, we would thus prefer to place these two geographic regions adjacent to the clusters suggested above.

Cluster IA consists of 60 species, including four endemic to the cluster (*Labyrinthocyathus squiresi*, *L. facetus*, *Cryptotrochus carolinensis* and *Thalamophyllia gombergi*), and zoogeographically corresponds to the warm temperate western Atlantic Carolinian Province. It is thus a fairly well differentiated group in both the cluster and ordination analyses, characterized by having species that have a relatively high amphi-Atlantic component, few species occurring deeper than 1000 m, and a higher than average percentage of free-living species. It corresponds to Briggs' (1974) Atlantic coast section of the Carolinian Province, pattern 4 of Cairns (1979) based on deep-water corals, and pattern 5 (Carolinian) of Cairns (2000) based on shallow-water azooxanthellates. Bermuda, having a deep-water fauna of only 19 species, does not cluster with any other region, but does fall within the domain of the western Atlantic (I) supercluster by standard clustering and ordination analyses. It has no distinctive endemic species, but does have an unusually high percentage of amphi-Atlantic and cosmopolitan species at the expense of a relatively low number of endemic western Atlantic species. In fact, one species, *Caryophyllia sarsiae*, is known only from the warm temperate eastern Atlantic and Bermuda, an indication of the independent nature of the Bermudian deep-water corals from the tropical western Atlantic. These faunistic components ally it closer to cluster IA (warm temperate w. Atlantic) and even to the North and eastern Atlantic superclusters instead of the tropical western Atlantic region (e.g., cluster IB), where conventional shallow-water zoogeography would place it.

The eastern Atlantic supercluster (III), consisting of 68 species, is also well supported by 26 species (38%) endemic to the region, although it is less well defined than the western Atlantic supercluster. Zoogeographically, it extends from the Faroe Islands to the Gulf of Guinea and thus includes the cold and warm temperate and tropical eastern Atlantic realms. As indicated in Table 4, this supercluster has a disproportionately high number of amphi-Atlantic and cosmopolitan species as well as a higher than average number of deep-living species, a phenomenon discussed below. Although conventional zoogeography (Briggs, 1974) based on shallow water fauna would indicate two major provinces within this supercluster, a warm temperate Lusitanian and a tropical West African, the cluster and ordination analyses suggest four clusters and several outliers.

The largest cluster (IIIA), consisting of 62 species, extends from Great Britain to Senegal, including the Azores and Madeira; Briggs (1974) calls this cluster the Lusitanian Province. Statistically, it resembles the larger eastern Atlantic supercluster in which it sits, having high components of amphi-Atlantic, cosmopolitan, and deep-living species, but also having four species endemic to the cluster, three of which are endemic to the Azores: *Leptopsammia formosa*, *Caryophyllia alberti* and *C. foresti*. Although the Azores lie far off the coast of Europe, this region strongly clusters within the Lusitanian region, as do most Azorean shallow water organisms (Briggs, 1974). The region consisting of the Shetland, Orkney and Faroe Islands, including 19 species, is an outlier, not strongly clustering with any other; it clusters most closely to the Lusitanian region, but aspects of its high component of amphi-Atlantic, flabellid, and unattached species, as well as its transitional placement between the North and Eastern Atlantic superclusters on the ordination analysis and its position in the cold temperate region, favours an interpretation that this region belongs with the North Atlantic supercluster but forms a faunistic transition between the two superclusters. It has one endemic species: *L. brittanica*. The 21 species occurring in the Mediterranean do not strongly cluster with any other regions, but only loosely to the Lusitanian cluster, with which it is normally associated (Briggs, 1974). The Mediterranean has no endemic deep-water species, the fauna being characterized as having a very high cosmopolitan component, few deep-living species, and a disproportionately high number of colonial and attached species (Table 4). Its central, but isolated, position on the ordination graph suggests that it might have served as a Tethyan centre of distribution for the surrounding regions.

It is difficult to understand why the cluster and ordination analyses distinguished cluster IIIB: the 28 species found on the seamounts south of the Azores and off Portugal. Whereas this cluster has one endemic species (*Paracyathus arcuatus*), the remaining 27 species are all found in the Lusitanian cluster (IIIA) and the statistics for this small cluster are unexceptional. It is suggested that this cluster be combined with the larger Lusitanian cluster (IIIA).

Cluster IIID consists of 21 species found off the Canary and Cape Verde Islands, the former island group in the warm temperate region, the latter at the northern edge of the tropical region. Although this cluster has no endemic species and is quite similar to the Lusitanian cluster in species composition, it is distinguished by having a very high cosmopolitan component, a high proportion of colonial species, and a low proportion of deep-living species. It also contains the species *Cladopsammia manuelensis*, which does not occur in the Lusitanian region.

Cluster IIIC consists of 24 species and represents the tropical eastern Atlantic realm, although very few if any species are known from a large part of this region from the continental slope off Dakar to the Gulf of Guinea. In general, this cluster falls near that of the Canary and Cape Verde Islands (IIID), but contains one endemic deep-water species (*L. chevalieri*) and two species not known from the Lusitanian region or Canary/Cape Verde regions: *Balanophyllia floridana* and *Ceratotrochus magnaghii*. This cluster has a high component of cosmopolitan species, and a high percentage of colonial and dendrophylliid species.

Whereas it is clear that both the western and eastern Atlantic superclusters (I and III, respectively) are each distinguished by a significant number of endemic species (Table 4), the boreal North Atlantic supercluster (II) contains only one endemic species (*Vaughanella margaritata*), but is nonetheless fairly well differentiated from the other two clusters. Instead of endemics, this region, which oceanographically corresponds to the cold temperate North Atlantic, is characterized by having an unusually high percentage of widespread (cosmopolitan) species, as well as all four contiguous amphi-Atlantic species; deep-living species; and an unusually low percentage of attached species. For instance, cluster IIA (cold temperate U.S. and Canada) consists of 47% widespread species and 28% amphi-Atlantic species, both numbers well above the North Atlantic average. Furthermore, this region is characterized by having many deep-living species, 71% of the 17 species from cluster IIA occurring deeper than 1000 m. Taxonomically, the fauna is heavily weighted towards flabellids (i.e., *Flabellum alabastrum*, *F. angulare* and *F. macandrewi*, three of the four contiguous amphi-Atlantic species) and the dendrophylliids are poorly represented. Cluster IIB (cold temperate northeastern Atlantic) is characterized by an even greater proportion of widespread species (i.e., 55%) and a relatively high amphi-Atlantic component, although including only two of the contiguous amphi-Atlantic species. This cluster also contains a higher than average number of deep-living species and a paucity of dendrophylliids. Although the Arctic region was considered to be an outlier in the cluster and ordination analyses, probably based on its low number of species, Table 4 indicates it is related to the other cold temperate regions of cluster II, being composed of a high percentage of widespread and contiguous amphi-Atlantic, deep-living, solitary species.

The fauna of St. Peter and Paul Rocks, two very small islands just north of the equator on the axis of the mid-Atlantic Ridge, is very poorly known; Briggs (1974) does not even commit to a zoogeographic placement. Because only six deep-water corals are known from these islands, it was treated as an outlier in the cluster analysis and not plotted in the ordination. Befitting its intermediate geographic position, its coral fauna also has largely intermediate affinities (50% cosmopolitan and 33% amphi-Atlantic), but one species, *Madrepora carolina*, indicates a western Atlantic affinity; its larvae are presumably transported by the Equatorial Counter Current.

Only three species are known from the New England Seamounts (*Desmophyllum dianthus*, *Solenosmilia variabilis* and *Enallopsammia rostrata*), not enough to formulate a meaningful analysis. All three species are deep-living cosmopolitan species, suggesting an affinity with the North Atlantic supercluster rather than the warm temperate region.

Amphi-Atlantic Depth Distributions

Zibrowius (1980: 221) noted that in general, the upper bathyal (200-1000 m) azooxanthellate corals of the eastern Atlantic were often found in deeper water than those in the western Atlantic, lamenting the fact that he had to trawl deeper in the east to get the same species Cairns obtained at shallower depths in the west. He hypothesized that each coral species has a temperature range preference and that corresponding isotherms were deeper in the eastern Atlantic. This is a logical explanation with which we agree, but we would like to elaborate. Azooxanthellate corals need not only the proper temperature range, but also a certain minimum amount of space (substrate) to populate and within which to evolve. The largest, most diverse, and most favorable region in the western Atlantic for deep-water azooxanthellate corals is the tropical Caribbean, some subregions of this province (e.g., Cuba, Lesser Antilles) having as many as 70-71 deep-water species. The warm temperate western Atlantic region is geographically compressed, does not have many offshore islands, and has a much smaller species diversity. In the eastern Atlantic, the region having the greatest geographic complexity, volume and species diversity is the warm temperate Lusitanian Province. In this case, the eastern Atlantic tropical region is latitudinally constricted, has narrow shelf and slopes, few islands and seamounts, and is thus less favourable for marine life, including deep-water azooxanthellate corals. Thus, if one were to consider an amphi-Atlantic species that had a temperature preference for 5-10°C, one might expect to find it at 500-1200 m in the tropical latitudes of the Antilles, but at 1300-1800 m in the warm temperate latitudes of the eastern Atlantic, the depths at which corresponding temperature ranges are found (Zibrowius, 1980). Indeed, among the 14 deep-water amphi-Atlantic Scleractinia, eight are known only from the tropical west and warm temperate east, in five cases the eastern populations occurring at significantly greater depths (no significant difference in one species); another eight species are known from the tropical and temperate west but only the warm temperate east, in four cases the eastern populations occurring at significantly greater depths (no significant difference in three species). Thus, among the 16 amphi-Atlantic species with asymmetric distributions, nine species were found at greater depths in the east, three at greater depths in the west, and there was no appreciable difference between east and west with the remaining four. The data in Table 4 also confirm that, in general, eastern Atlantic species tend to occur at greater depths than western Atlantic species.

CONCLUSIONS

The first step in conserving or managing a resource such as the deep-water corals of the North Atlantic, is to know the number of species, and their distribution and abundance within that region. Tables 1 and 2 address that issue. But knowing other biological and zoogeographic information about the species and species clusters to which they belong may also allow some predictions about the stability or vulnerability of the fauna of various regions. These factors are quantified in Table 4. For instance, one would assume that a fauna having a higher-than-average percentage of cosmopolitan and amphi-Atlantic species and a low percentage of endemic species would constitute a more stable fauna, since the extirpation of one or more species from such a region would more easily be replaced from an adjacent region, and there would be a lesser chance for an endemic to go extinct. Likewise, faunas having a higher-than-average percentage of deep-water species (over 1000 m) would be more stable than one having numerous shallow-water species. If these assumptions are correct, the most stable region in the North Atlantic is cluster IIA, specifically the cold temperate U.S. and Canada (region CTC), which has 60% cosmopolitan, 30% amphi-Atlantic and 10% endemic components, and 90% of whose species occur deeper than 1000 m. Conversely, the most vulnerable region or cluster appears to be IB (insular western Atlantic), which has a 27% endemic component, and only 20% of whose species occur deeper than 1000 m. This is somewhat ironic in that this most vulnerable region has the highest species diversity in the North Atlantic, and the CTC one of the smallest, but this may be due to a combination of a faster adaptive radiation of deep-water corals in the tropical western Atlantic than in the more temperate regions, and little or no decrease by extinction as yet.

ACKNOWLEDGEMENTS

We would like to thank staff illustrator Molly Ryan for drafting Figures 1-5, and John Dawson and Charles R.C. Sheppard for reviewing the manuscript and offering helpful comments.

REFERENCES

Bray, J.R. and J.T. Curtis. 1957. An ordination of the upland forest communities of southern Wisconsin, Ecological Monographs 27: 325-349.

Briggs, J.C. 1974. Marine Zoogeography. McGraw-Hill Book Co., New York, 475 pp.

Cairns, S.D. 1979. The deep-water Scleractinia of the Caribbean Sea and adjacent waters, Studies on the fauna of Curaçao and other Caribbean Islands 180, 341 pp.

Cairns, S.D. 1981. Marine flora and fauna of the northeastern United States. Scleractinia, NOAA Technical Report NMFS Circular 438, 14 pp.

Cairns, S.D. 1992. Worldwide distribution of the Stylasteridae (Cnidaria, Hydrozoa), Scientia Marina 56(2-3): 125-130.

Cairns, S.D. 1999. Species richness of Recent Scleractinia, Atoll Research Bulletin 459: 1-12.

Cairns, S.D. 2000. A revision of the shallow-water azooxanthellate Scleractinia of the western Atlantic, Studies of the Natural History of the Caribbean

75: 231 pp.
Clarke, K.R. and R.M. Warwick. 1994. Change in Marine Communities, An Approach to Statistical Analysis and Interpretation, Natural Environment Research Council, UK, 144 pp.
Cormack, R.M. 1971. A review of classification, Journal of the Royal Statistical Society 134: 321-367. [not seen]
Ekman, S. 1953. Zoogeography of the Sea. Sidgwick and Jackson, London, 417 pp.
Field, J.G., K.R. Clarke, and R. M. Warwick. 1982. A practical strategy for analyzing multispecies distribution patterns, Marine Ecology – Progress Series 8: 37-52.
Hartigan, J.A. 1975. Clustering Algorithms. Wiley, New York, 351 pp.
Ketchum, J.T. and H. Reyes-Bonilla. 1997. Biogeography of hermatypic corals of the Archipelago Revillagigedo, Mexico. Proceedings of the 8th International Coral Reef Symposium, Guam 1: 471-476.
McCune, B. and M.J. Mefford. 1999. PC-ORD. Multivariate Analysis of Ecological Data, Version 4. MjM Software Design, Gleneden Beach, Oregon, 237 pp.
Reyes-Bonilla, H. and A. López-Pérez. 1998. Biogeography of the stony corals (Scleractinia) of the Mexican Pacific, Ciencias Marinas 24(2): 211-224.
Roberts, D.W. 1986. Ordination on the basis of fuzzy set theory, Vegetation 66: 123-131.
Sánchez, J.A. 1999. Black coral-octocoral distribution patterns on Imelda Bank, a deep-water reef, Colombia, Caribbean Sea, Bulletin of Marine Science 65(1): 215-225.
Sánchez, J.A., S. Zea, and J.M. Díaz. 1998. Patterns of octocoral and black coral distribution in the oceanic barrier reef-complex of Providencia, Southwestern Caribbean, Caribbean Journal of Science 34(3-4): 250-264.
Sheppard, C.R.C. 1997. Biodiversity patterns in Indian Ocean corals, and effects of taxonomic error in data, Biodiversity and Conservation 7: 847-868.
Sheppard, C.R.C. 1999. Corals of Chagos, and the biogeographical role of Chagos in the Indian Ocean. In C.R.C. Sheppard and M.R.D. Seaward (eds), Ecology of the Chagos Archipelago. Occasional Publications of the Linnean Society of London, vol. 2, Burlington House, Picadilly, London, pp. 53-66.
Sneath, P.H.A. and R.R. Sokal. 1973. Numerical Taxonomy: The Principles and Practice of Numerical Classification. Freeman, San Francisco, 573 pp.
Tyler, A. and H. Zibrowius. 1992. Submersible observations of the invertebrate fauna on the continental slope southwest of Ireland (NE Atlantic Ocean), Oceanologica Acta 15(2): 211-226.
Veron, J.E.N. 1995. Corals in Space and Time; the biogeography and evolution of the Scleractinia. University of New South Wales Press, Sydney, 321 pp.

Wallace, C. 1999. Staghorn Corals of the World, a revision of the genus *Acropora* (Scleractinia; Astrocoeniina; Acroporidae) worldwide, with emphasis on morphology, phylogeny and biogeography. CSIRO Publishing, Collingwood, Vic., Australia, 421 pp.

Zibrowius, H. 1980. Les scléractiniaires de la Méditerranée et de l'Atlantique nord-oriental, Mémoires de l'Institut Océanographique, Monaco 11: 84 pp.

Zibrowius, H. 1983. Nouvelles données sur la distribution de quelques Scléractiniaires "Méditerranéens" à l'est et à l'ouest de détroit de Gibraltar, Commission internationale pour l'exploration scientifique de la mer Méditerranée. Rapports et procès-verbaux des réunions, pp. 207-209.

Zibrowius, H. 1985. Scléractiniaires bathyaux et abyssaux de l'Atlantique nord-oriental; campagnes BIOGAS (POLYGAS) et INCAL. In L. Laubier and C. Monniot (eds), Peuplements profunds du Golfe de Gascogne. IFREMER, France, pp. 311-324.

Zibrowius, H. 1992. The Scleractinia (Cnidaria, Anthozoa) of the BIOFAR project. Symposium on marine biology and oceanography of the Faroe Islands, Tórshavn/Faroe Islands, 16-20 September 1991. Norðurlandahúsið ársrit, Tórshavn, 1991-92: 78-79.

OBSERVATIONS ON THE OCCURRENCE AND HABITAT PREFERENCE OF CORALS IN ATLANTIC CANADA

Kevin MacIsaac[1], Cynthia Bourbonnais[1], Ellen Kenchington[1], Donald Gordon Jr.[1], and Susan Gass[2]
[1] Fisheries and Oceans Canada, Marine Environmental Sciences Division, Bedford Institute of Oceanography, Dartmouth, Nova Scotia
[2] Ecology Action Centre, Halifax, Nova Scotia

SUMMARY
The distribution, abundance and habitat of deep-sea corals in Atlantic Canada have been investigated using several opportunistic approaches, including benthic surveys, observations from experimental areas, and incidental bycatch in groundfish surveys. These efforts included sites ranging from the Northeast Channel to the northern Labrador Shelf. Most information on habitat was collected using new gear developed at the Bedford Institute of Oceanography and has allowed direct observations of coral. A total of 15 species and species groups were observed, with details of location and species associated with each coral presented briefly. Corals were found at all areas surveyed and results are in general agreement with recent anecdotal observations, though they expand the ranges of several species. The highest densities of corals were observed on Romeys Peak, at the mouth of the Northeast Channel, in areas with high current velocities on elevated iceberg scours and pits. The highest diversity of coral was observed at the Gully and the Stone Fence, where surveys covered a wide variety of habitats.

INTRODUCTION

Popular use of the word "coral" has evolved over the past decade to include animals beyond the familiar reef-building corals of the tropics. Deep-sea corals are found in all of the world's oceans and the Mediterranean Sea. The deep water off Atlantic Canada supports approximately 35 species of coral (Breeze et al., 1997): seven Alcyonacea (soft corals), ten Gorgonacea (horny corals), eight Pennatulacea (sea pens) and ten Scleractinia (stony corals). Most appear to be found in canyons along the shelf edge or in deep channels between fishing banks. Most information concerning the distribution of coral off Nova Scotia is dated (i.e., Verrill, 1885; Whiteaves, 1901; Deichmann, 1936), inferred from other areas (i.e., Hecker et al.,1980; Cimberg et al., 1981) or is anecdotal in nature (Breeze et al., 1997). Interviews with local fishers have indicated reductions in corals and changes in the ocean floor in several areas between Georges Bank and the Laurentian Channel (Breeze et al., 1997). There is concern that human activities in and around submarine canyons and along the edge of the continental shelf may affect habitat and communities that are not yet fully understood. Little is known about the potential role of corals in deep-sea ecosystems, as habitat for other species, or how they may possibly be linked to shallower shelf ecosystems (i.e., commercial fisheries). Research on corals is necessary to increase scientific knowledge and to help develop sound scientific advice on habitat and fisheries management issues related to corals.

Since 1997, the Department of Fisheries and Oceans (DFO) has been collecting qualitative and quantitative information on corals and coral habitat in Atlantic Canada using a variety of different approaches. A series of benthic surveys has been conducted in areas reported to have (had) high coral densities. Surveys have been planned with input from industry, government managers and local non-government organizations, and carried out opportunistically while conducting other benthic habitat programs. In addition to planned surveys, the report includes observations from two experimental sites and data on incidental coral bycatch from DFO groundfish surveys.

This report presents some preliminary results on distribution, abundance, habitat and associated species, and is intended to be descriptive. These data are nevertheless considered to be of immediate value because of the growing interest in corals and the limited biological and ecological information available.

MATERIALS AND METHODS

Areas Investigated

Benthic Surveys

Areas surveyed extended from the Stone Fence to the Northeast Channel (Figure 1). Surveyed in 1997, 1999 and 2000, the Gully is a large submarine canyon located between Sable Island Bank and Banquereau. Specifics of its ecosystem and environment have been reviewed in detail (Harrison and Fenton, 1998). The Gully cuts across the shelf and down the slope to depths over 2000 m, with substrates ranging from fine, well-sorted sand to gravel areas to slump deposits with rock outcrops and sheer cliffs (Lawrence et al., 1985; Energy, Mines and Resources, 1986; Fader et al., 1998). Temperature and salinity vary with both season and depth, and range from about 2 to 15°C and 30 to 35 psu (Petrie et al., 1998). Current direction is predicted to be complex with gyre-like features and movement up- and down-slope (Petrie et al., 1998).

Surveyed in 1999, Haldimand Canyon is a prominent feature off Banquereau with substrates consisting of mainly fine, well-sorted sand with shell material in some ridges and occasional rock outcrops in deeper areas (Lawrence et al., 1985; Energy, Mines and Resources, 1986). Temperature ranges from about 2 to 15°C and salinity from 31 to 35 psu (Petrie et al., 1996). Mean current direction below 80 m is predicted to be a combination of incoming water from off-shelf and along-shelf from east to west (Gregory and Bussard, 1996).

Surveyed in 1999, the Stone Fence is located at the easternmost edge of the Scotian Shelf bordering the Laurentian Channel (Figure 1). The substrate

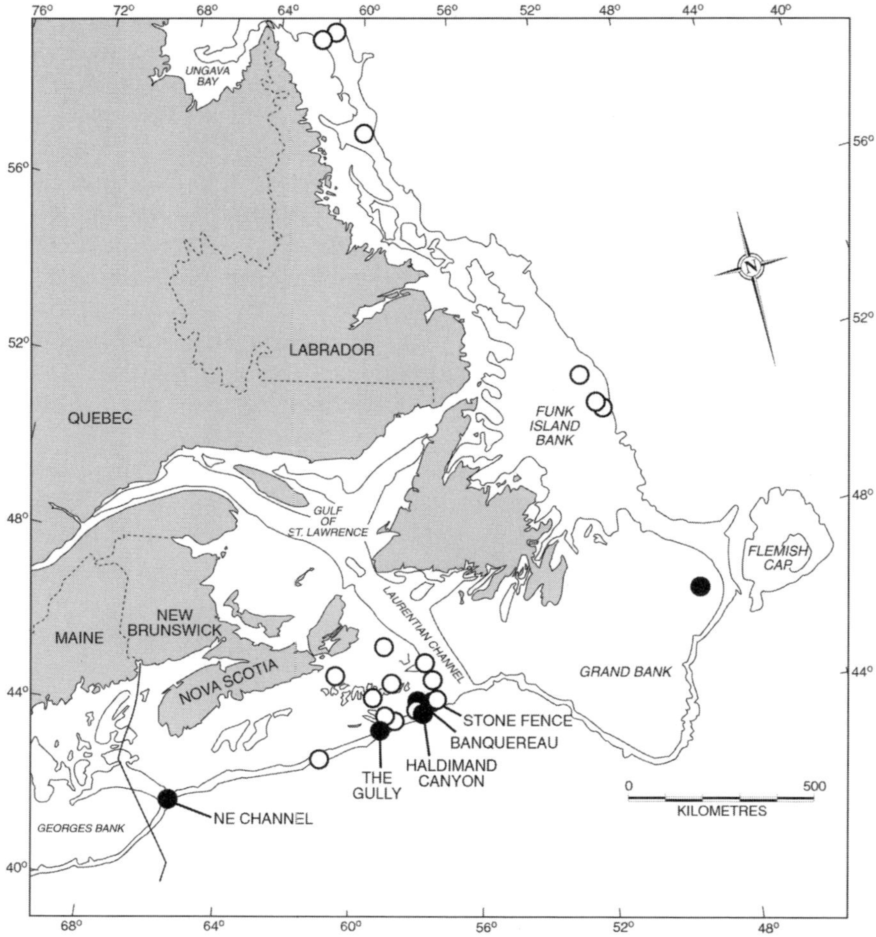

Figure 1. Approximate locations of opportunistic benthic surveys and experimental sites (closed circles), and groundfish trawls with coral bycatch (open circles)

consists of abundant cobble and boulders with occasional sand bedforms and old iceberg furrows (Lawrence et al., 1985; Energy, Mines and Resources, 1986). Temperature ranges from about 2 to 15°C and salinity from 31 to 35 psu (Petrie et al., 1996). Mean current direction below 80 m is predicted to be complex, with movement back and forth along the shelf as well as from off-shelf (Gregory and Bussard, 1996).

Surveyed in 2000, the Northeast Channel is located between Browns Bank and Georges Bank (Figure 1). The substrate is composed of glacial till with moraines, the surface covered with iceberg pits and furrows with large exposed boulders around the rims (Lawrence et al., 1985; Energy, Mines and Resources, 1986; Fader et al., 1988). Some areas have sand in transport from the high velocity currents in and around the channel. Temperature ranges from about 4 to 15°C and salinity from 32 to 35 psu (Petrie et al., 1996). There is a cyclic pattern in current direction and velocity in this area, with strong currents moving up-channel during high tide, down-channel during low tide, and cross-channel during periods of slack tide (Greenberg, 1983; Valentine and Lough, 1991).

Study Sites

The study site on the Grand Bank was surveyed from 1993 to 1995. It is located in relatively deep water (120 to 146 m) with surficial sediments composed of a moderately to well-sorted, fine- to medium-grained sand (Lawrence et al., 1985; Energy, Mines and Resources, 1986). Temperature ranges from about 0° to 10°C and salinity from 32 to 34 psu. Mean current in this area is predicted to be complex, with movement into the area from off the shelf (Gregory et al., 1996). The study site on Banquereau was surveyed from 1998 to 2000. It is a relatively flat, featureless seabed at a depth of 60 to 70 m with fine-grained, well-sorted sand and relatively dense concentrations of shell material (Figure 1). The site is near Haldimand Canyon, so information on temperature, salinity and current are expected to be fairly similar (Lawrence et al., 1985; Energy, Mines and Resources, 1986; Gregory and Bussard, 1996; Petrie et al., 1996).

Groundfish Surveys

Groundfish surveys were conducted from the coast of Labrador to the Scotian Shelf (Figure 1). These areas contain a variety of habitats including a broad range of substrates (sands and gravels, muddy sands and gravels, basin muds, silts and glacial drift), depths (47 to 1179 m) and hydrographic conditions. Only trawls that contained coral as incidental bycatch during 1999 and 2000 were included.

Types of Gear

CAMPOD

Photo and video records (S-VHS and digital format) were collected using video and photographic cameras mounted on an aluminum tripod platform called CAMPOD (Milligan et al., 1998; see Plate 1, top left). Survey transects consisted of a series of discrete stations of variable depth and duration, to a maximum depth of approximately 450 m. A time code (GMT) was recorded on the videotape along with ship's navigation information. Ship position was calculated using a Differential Global Positioning System. Gear position was estimated by using a five-metre offset from ship position, to coincide with the tip of the boom used for lowering equipment (McKeown and Gordon, 1997). Video provided qualitative information on substrate and larger features such as outcrops and bedforms. Presence of larger organisms was noted

Table 1. Corals recorded during recent DFO surveys (1993-2000). Nomenclature as presented in Breeze et al. (1997)

Coral species	Location observed/collected	Gear
Subclass Octocorallia		
Order Alcyonacea		
Anthomastus purpureus	Stone Fence	Video, photo
Alcyonacea-type 1	Haldimand Canyon, Stone Fence	Video, photo
Alcyonacea-type 2	Gully, Haldimand Canyon, Stone Fence	Video, photo
Gersemia rubiformis	Banquereau, Grand Bank	Video, photo, grab, sled
Order Gorgonacea		
Paragorgia arborea	NE Channel, Gully, Stone Fence, N. Labrador	Video, trawl
Acanthogorgia armata	Gully, Stone Fence	Video, photo, grab, trawl
Keratoisis ornata	Gully, Stone Fence	Video, photo
Gorgonacea-type 1	S. Labrador	Trawl
Gorgonacea-type 2	Stone Fence	Video
Primnoa resedaeformis	NE Channel, Gully, Stone Fence, Labrador	Video, grab, trawl
Acanella arbuscula	Gully, E. Scotian Shelf	Grab, trawl
Order Pennatulacea		
Balticina finmarchica	Gully, Stone Fence	Video
Pennatula sp.	Gully, Haldimand Canyon	Video, photo
Undetermined sea pen	Stone Fence	Video, photo
Subclass Hexacorallia		
Order Scleractinia		
Flabellum alabastrum	NE Scotian Shelf	Trawl

with qualitative comments concerning abundance. Photos provided quantitative counts or rankings of macrofauna and details of substrate. All photographs have been digitized and are stored on compact disc.

Videograb

The videograb is a large benthic grab (0.5 m^2) equipped with video that can work to a maximum depth of about 450 m (Rowell et al., 1997). After samples of sediment and biota were collected, total grab volume was measured and then washed through a 1 mm sieve. The remaining material (sediment and biota) was fixed with 4% buffered formaldehyde for transport to a laboratory where all macrofauna were identified to lowest taxonomic level possible, counted, weighed (wet), and preserved in 70% ethanol. This was a central piece of equipment at experimental sites (Kenchington et al., 2001), but was only used in a limited way at survey locations to selectively retrieve live and relatively undisturbed specimens of coral.

Epibenthic Sled

The epibenthic sled is a modified dredge with a 0.33 m opening and a collection box with a mesh size of 1 cm that was designed to sample the upper few centimetres of the seafloor (Rowell et al., 1997). Collected material was washed through a 1 mm sieve and all macrofauna identified to lowest taxonomic level possible, counted and weighed (wet). This was a central piece of equipment used at the Grand Bank experimental site (Prena et al., 1999), but was not used elsewhere.

Groundfish Trawls

Annual groundfish trawl surveys conducted by DFO not only provide information on fish species and biomass, but also on invertebrate bycatch, location, depth, temperature, salinity and oxygen (Halliday and Koeller, 1981). Beginning in 1999, incidental coral bycatch was retained and frozen onboard for transport ashore to a laboratory, where specimens were thawed, weighed (wet), and closely examined for commensals. Commensals were quickly sorted, fixed in 4% buffered formaldehyde, and preserved in 70% ethanol for more detailed examination at a later date. Maximum dimensions (height and width) of each specimen were measured and recorded, and a sub-sample of tissue was collected, fixed in 4% buffered formaldehyde, and preserved in 70% ethanol. Specimens were then dried to a constant weight and recorded.

RESULTS

Benthic surveys were carried out at the Northeast Channel, the Gully, Haldimand Canyon and the Stone Fence. Five videograbs, over 250 minutes of video and 137 photos were collected in the immediate vicinity of coral; these methods produced information on 12 species and species groups (data from the 2000 surveys not fully analyzed). Experimental sites on Banquereau and Grand Bank provided observations on one species of soft coral, *Gersemia rubiformis* (Prena et al., 1999; Kenchington et al., 2001). In addition, a total of 19 groundfish trawls from Newfoundland/Labrador and the Scotian Shelf collected 41 specimens of coral including eight species. Overall, information was collected on 15 species and species groups (Table 1). Taxonomy and nomenclature are as presented in Breeze et al. (1997).

The coral fauna

The Alcyonaceans

Anthomastus purpureus (Koren & Danielssen, 1883)

This species was recorded in video footage and photo at the Stone Fence at a depth of 398 m attached to cobble in an area of abundant gravel. Other co-occurring corals in the same general area include Alcyonacea-type 2, *Acanthogorgia armata* and *Keratoisis ornata*. No commensals were apparent in video or photos, but unidentified sponges (Porifera) and actinarian anemones were abundant on surrounding hard substrata, and redfish (*Sebastes* sp.) were common in the immediate area of the coral.

Anthomastus purpureus has been reported from Newfoundland to the Caribbean at depths between 200 and 2600 m. Present records of this genus occur within the historical range of depth and latitude (Deichmann, 1936).

Alcyonacea-type 1 (*Gersemia rubiformis* or *Alcyonium digitatum*)

Commonly referred to as sea cauliflower, strawberries, red soft coral (G. rubiformis) or dead man's fingers (A. digitatum), this species group was recorded in video footage and photos at Haldimand Canyon and the Stone Fence (Plate 1, centre left). G. rubiformis and A. digitatum may look similar in photos and video, and depths of distribution overlap off Nova Scotia, so no attempt was made to separate the two, unless samples were collected for verification. It was observed attached to shell hash in areas of well-sorted sand and attached to cobble in gravelly areas. It could be moderately abundant in shallow, sandy locations with abundant shell hash. This species group had the shallowest distribution observed, with a depth of occurrence of 67-94 m. The only other coral occurring in the same general area as Alcyonacea-type 1 was Alcyonacea-type 2. Commensals were not apparent on colonies in video or photos at either location. At Haldimand Canyon, there was an abundance of co-occurring epifauna, including sand dollars (Echinarachnius parma), brittle stars (Ophiura) and buccinid whelks. Colonies of thecate hydrozoans commonly occupied the same or similar substrate. With increasing depth came a marked decrease in abundance of Alcyonacea-type 1, shell hash became rare, and the abundant epifauna disappeared. Stalked tunicates (Boltenia) became more common and large feeding aggregations of the purple sea cucumber (Cucumaria frondosa) were observed. No other corals were observed with Alcyonacea-type 1 in Haldimand Canyon. Undetermined flounder (Pleuronectiformes) and skate (Rajidae) were the only fish species. At the Stone Fence, Alcyonacea-type 1 was not as abundant. It was observed attached to hard substrate on a predominantly cobble bottom encrusted with varying amounts of coraline algae and thecate hydrozoa. Polychaete worm tubes (Onuphidae), Ophiura, sea urchins (Strongylocentrotus) and Boltenia were fairly common. A small, unidentified species of fish was occasionally seen resting on the irregular cobble bottom.

A. digitatum is reported from Nova Scotia to New England at depths between 10 and 170 m, and *G. rubiformis* from Greenland to New England at between 40 and 100 m. Present records occur within the historical ranges of depth and latitude for both species (Whiteaves, 1901; Deichmann, 1936; Miner, 1950).

Alcyonacea-type 2 (*Drifia glomerata*, *Duva florida* or *Gersemia fruticosa*)

Collectively referred to as sea cauliflower, strawberries or simply soft coral, this species group was the most abundant type of alcyonacean coral in the surveyed areas. *Drifia glomerata*, *Duva florida* and *Gersemia fruticosa* may look similar in photos and video, and depths of distribution overlap in Nova Scotia. All may have been observed during surveys, and species are known to occupy the same substrate, so no attempt was made to distinguish them. Recorded in video and photos from the Gully to the Stone Fence, it was observed attached to hard substrate in areas with varying amounts of soft sediment and a wide range of gravel and outcrops. Depth of occurrence ranged from 67 to 365 m. Alcyonacea-type 2 co-occurred in the same general area with *Anthomastus purpureus*, Alcyonacea-type 1, *Paragorgia arborea*, *Acanthogorgia armata*, *Keratoisis ornata*, Gorgonacea-type 1, *Primnoa resedaeformis*, *Balticina finmarchica*, *Pennatula* sp. and an undetermined white-coloured sea pen. Commensals included the basket star (*Gorgonocephalus*) and brittle stars (ophiuroids). An abundance of epifauna encrusted the surrounding hard substrate, including undetermined actinarian anemones, various sponges and polychaete worm tubes (Serpulidae). At deeper locations various ophiuroids occupied the intervening soft substrate, which included high densities of a mud-burrowing species (possibly *Amphiura*). Common fish species included *Sebastes*, longfin hake (*Urophycis chesteri*), undetermined grenadier (Macrouridae) and pollock (*Pollachius virens*).

Two varieties of Alcyonacea-type 2 were noticed: a blue (Plate 1, centre right) and a white form that seem to correspond to descriptions of *Gersemia fruticosa* and *D. florida*/*D. glomerata*, respectively. *Drifia glomerata* may be the shallowest of the three species at this latitude. *Duva florida* may have a distribution mode at about 400 m. *Gersemia fruticosa* appears restricted to depths greater than 2000 m off the United States, but occurs at 100 m in the Arctic (Deichmann, 1936; Hecker et al., 1980). Relative proportions of the three species cannot be estimated based on this information. Species are reported to occur at approximately 100 to 3000 m depth from Greenland to New England (Whiteaves, 1901; Deichmann, 1936; Miner, 1950; Hecker et al., 1980). Some of these records may be shallower than previously reported for species if they were separable.

Gersemia rubiformis (Ehrenberg, 1834)

Commonly known as sea cauliflower, strawberries or red soft coral, this species was recorded in video, photos and videograbs at the Banquereau experimental site and in videograbs and epibenthic sled at the Grand Bank experimental site (Prena et al., 1999; Kenchington et al., 2001). It was moderately abundant, attached to shell hash and occasional pebbles in areas of well-sorted sand. Along with Alcyonacea-type 2, this species was the shallowest in the surveyed areas, occurring at a depth of 60 to 140 m. Other coral species did not co-occur in the same general area as *G. rubiformis*, and *Gorgonocephalus* was the only species observed to be commensal. Dominant epifauna at both sites were the mobile echinoderms, including sand dollars *Echinarachnius parma*, brittle stars (*Ophiura sarsi*), sea urchins (*Strongylocentrotus pallidus*), *Gorgonocephalus* (at the Grand Bank site) and *Cucumaria frondosa* (at the Banquereau site). Fish species observed at the Grand Bank site included plaice (*Hippoglossoides platessoides*) and thorny skate (*Raja radiata*). Yellowtail flounder (*Limanda ferruginea*), various skates (Rajidae) and sculpin (Cottidae) were common on Banquereau. The Grand Bank collections of *G. rubiformis* are deeper than previously reported (Whiteaves, 1901; Deichmann, 1936; Miner, 1950).

The Gorgonaceans

Paragorgia arborea (Linnaeus, 1758)

This relatively familiar species is known locally as rubber trees, bubblegum coral and strawberry plants (Breeze et al., 1997). Though not abundant, this species was one of the most widespread corals at depths less than 450 m, and was recorded in video from the Northeast Channel to the Stone Fence and as incidental trawl bycatch along the coast of Labrador. It was observed attached to large cobble, boulders and near-vertical walls in areas with a mixture of sand, muddy sand, and gravel with occasional outcrops. Depth of occurrence was narrow, between 321 and 375 m. Other co-occurring corals in the same general area included Alcyonacea-type 2, *Acanthogorgia armata*, Gorgonacea-type 2 and *Primnoa resedaeformis*. There were no commensals apparent in video or on the collected specimens. An abundance of epifauna including undetermined actinarian anemones, various sponges, and serpulid worm tubes encrusted the surrounding hard substrate. *Sebastes* was the most common species of fish observed in video. In groundfish trawls the invertebrate bycatch accounted for most of the total catch, and was dominated by natant decapod shrimp (probably *Pandalus*), sponges and other undetermined invertebrates. The deep-water redfish (*Sebastes mentella*) and arctic eelpout (*Lycodes reticulatus*) dominated the fish catch.

Paragorgia arborea is distributed from Georges Bank to Greenland and across to Norway at depths between 200 and 1000 m. Present records occur within the reported range of depth, but trawl collections from Labrador now represent the northernmost collections of *P. arborea* on the east coast of North America (Whiteaves, 1901; Deichmann, 1936; Miner, 1950; Hecker et al., 1980; Tendal, 1992; Breeze et al., 1997).

Acanthogorgia armata Verrill, 1878

This small, salmon-coloured coral was recorded in video, photo and videograb, from the Gully to the Stone Fence and as incidental bycatch in groundfish trawls on the eastern Scotian Shelf around Banquereau and Misaine Bank. It was collected in areas reported to be a mixture of sand, muddy sand, and gravel of various sizes, where it was observed attached to pebbles or granules. Depth of occurrence ranged from 150 to 384 m, with dead specimens found at the shallower locations. No other corals were found at the shallow sites with dead colonies. Other corals found in the same general area with living colonies of *Acanthogorgia armata* include *Anthomastus purpureus*, Alcyonacea-type 2, *Paragorgia arborea*, *Keratoisis ornata*, Gorgonacea-type 2 and *Acanella arbuscula*. Commensals included ophiuroids, skeleton shrimp (caprellid amphipods), bryozoa, scalpellid barnacles, hydrozoa, polychaete worms (Oweniidae and possibly Terebellidae) and thin mats of an undetermined tissue-like material. The invertebrate bycatch in groundfish trawls was less than one percent of the total catch and was dominated by sea stars (Asteroidea), stone crab (*Lithoides maja*) and snow crab (*Chionoecetes opilio*). *Sebastes* dominated fish biomass.

This species has been reported in the northwest Atlantic from Newfoundland south past Georges Bank at a depth range between 400 and 2150 m (Whiteaves, 1901; Deichmann, 1936; Miner, 1950; Hecker et al., 1980). These collections are considerably shallower than previously recorded.

Keratoisis ornata Verrill, 1878

Commonly referred to as bamboo coral, gold-banded coral, or birch trees, this species was recorded in video and photos from the Gully and the Stone Fence. It was observed at depths between 382 and 443 m in areas reported to be a mixture of sand with gravel, and muddy sand with gravel and occasional outcrops. It is not abundant at these locations and depths. Other corals co-occurring in the same general area include *Anthomastus purpureus*, Alcyonacea-type 2, *Acanthogorgia armata* and *Primnoa resedaeformis*. There were no commensals apparent in video or photos. An abundance of epifauna including undetermined actinarian anemones, various sponges, and serpulid worm tubes encrusted the surrounding hard substrate. Various ophiuroids occupied the intervening soft substrate, including high densities of a burrowing species at deeper, muddy/clay sites (possibly *Amphiura*). Fish observations included *Sebastes*, grenadier (Macrouridae) and unidentified gadoids (cod-like species).

This species is reported off Nova Scotia at a depth between 300 and 600 m, and at 1700 m off the West Indies. Present records occur within the historical range of depth and latitude (Whiteaves, 1901; Deichmann, 1936; Miner, 1950; Breeze et al., 1997).

Gorgonacea-type 1

This species was collected from incidental trawl bycatch along the coast of Labrador (Plate 1, lower right). Gorgonacea-type 1 could not be identified confidently based on the current list of species expected (Breeze et al., 1997) and is therefore considered a new record for the region. Information on substrate in areas of trawl collections is being compiled by DFO in Newfoundland. Gorgonacea-type 1 was the deepest occurring species in our survey, at depths ranging from 425 to 1179 m. It does not appear to be abundant, and was not collected with any other coral species. Commensals included feather stars (crinoids), ophiuroids, ctenostome bryozoa, *Gorgonocephalus*, thecate hydrozoa, colonial zoanthid anemones (*Epizoanthus*), hormathiid anemones, and thin mats of an undetermined tissue-like material. Invertebrate bycatch accounted for nearly half of the total catch in groundfish trawls, and was dominated by sponges, other undetermined invertebrates, octopus (Octopodidae) and Scyphozoa (it is unknown whether these were deep-sea benthic scyphozoa or common pelagic jellyfish). Turbot (*Reinhardtius hippoglossoides*), *Sebastes mentella* and roughhead grenadier (*Macrourus berglax*) dominated fish biomass.

Gorgonacea-type 2

This species was observed from video at the Stone Fence at a depth of 300 m attached to cobble in an area of abundant gravel, but was not common. Other corals found in the same general area include Alcyonacea-type 2, *Paragorgia arborea, Acanthogorgia armata* and *Primnoa resedaeformis*. The only commensal observed was a solitary feather star (Crinoidea). The redfish, *Sebastes*, was observed in video.

Primnoa resedaeformis (Gunnerus, 1763)

Primnoa resedaeformis was one of the most widespread and abundant species surveyed, and was recorded on video from the Northeast Channel to the Stone Fence, in videograbs from the Northeast Channel, and as incidental bycatch in groundfish trawls along the coast of Labrador (Plate 1, lower left and upper right). Collections were in areas of sand or muddy sand with gravel of various sizes, where it was observed attached to large cobbles and boulders. This species was particularly abundant on Romeys Peak in the Northeast Channel, where high densities were observed associated with iceberg furrows and pits, features with an abundance of large cobble and boulders elevated slightly above the surrounding seabed. Depth of occurrence ranged from 321 to 425 m. Other corals co-occurring in the same general area include Alcyonacea-type 2, *Paragorgia arborea*, *Acanthogorgia armata*, *Keratoisis ornata* and Gorgonacea-type 2. Commensals included decapod shrimp (*Pandalus propinquus*), thecate hydrozoa (especially on dead branches), anascan and cyclostome bryozoa, polychaete worms (Eunicidae), *Gorgonocephalus*, caprellid amphipods, scalpellid barnacles, thin mats of undetermined tissue, and sponges and ophiuroids around the base of colonies. An abundance of epifauna including the anemones *Bolocera tuediae* and *Actinauge verrillii*, other undetermined actinarian anemones, various sponges, and serpulid worm tubes encrusted the surrounding hard substrate. The fauna occupying the intervening soft substrate was occasionally abundant, including dense mats of filter-feeding ophiuroids (*Ophiacantha abyssicola*) on Romeys Peak. *Sebastes*, pollock (*Pollachius virens*) and cod (*Gadus morhua*) were common species of fish observed in video. Invertebrate bycatch accounted for nearly all of the biomass in groundfish trawls, and was dominated by undetermined sponges and anemones. *Sebastes mentella*, *Reinhardtius hippoglossoides* and myctophids (pelagic lantern fish probably caught during set/haul) dominated fish biomass.

Primnoa resedaeformis is reported from Greenland south past Georges Bank at depths between 100 and 600 m. These records occur within the historical range of depth and latitude (Whiteaves, 1901; Deichmann, 1936; Miner, 1950; Hecker et al., 1980; Breeze et al., 1997).

Acanella arbuscula (Johnson, 1862)

Colonies of this bright orange coral were recorded from videograb at the Gully and as incidental bycatch in groundfish trawls along the eastern Scotian Shelf in the area around Banquereau and Misaine Bank, but the species is not abundant. It was observed anchored in soft, muddy sand, in an area with sparse cobble and pebble. Depth of occurrence ranged from 111 to 384 m, with dead specimens at shallower locations only. No other species of coral were found at the shallow locations where the dead specimens were collected, but other corals in the same general area with living colonies of *A. arbuscula* include *Acanthogorgia armata* and *Flabellum alabastrum*. Commensals included *Gorgonocephalus*, other ophiuroids, caprellid amphipods, various anascan and cyclostome bryozoa, and thin mats of an undetermined tissue-like material. The invertebrate bycatch in groundfish trawls containing living *A. arbuscula* was insignificant, consisting of small amounts of mostly *Pandalus*, *Chionoecetes opilio* and *Lithoides maja*. *Sebastes* and haddock (*Melanogrammus aeglefinus*) dominated trawl biomass. In areas with dead colonies the invertebrate bycatch accounted for nearly one quarter of the total trawl catch, and was still dominated by the same species, but the fish assemblage differed and was dominated by *Hippoglossoides platessoides*, *Raja radiata* and *Sebastes*.

Acanella arbuscula has been reported on soft substrates from Newfoundland to Chesapeake Bay at depths of about 400 to 3500 m. It has been reported to be abundant in deep water around the banks of Nova Scotia and Newfoundland usually at about 600 to 800 m (Whiteaves, 1901; Deichmann, 1936; Miner, 1950; Hecker et al., 1980). These collections are considerably shallower than previously recorded.

The Pennatulaceans

Balticina finmarchica (Sars, 1851)

This tall, slender sea pen was tentatively identified from video footage at the Gully and Stone Fence, where it appeared moderately abundant in localized areas. It was anchored in soft (muddy) bottom with occasional gravel and outcrops at depths between 255 and 443 m. Other corals co-occurring in the same general area include Alcyonacea-type 2 and an undetermined white-coloured sea pen. Commensals were not apparent in video or photos and the only other common epifauna were burrowing anemones. *Sebastes* was common, with occasional *Pollachius virens*, grenadier (Macrouridae), flounder (Pleuronectidae) and ocean pout (*Macrozoarces americanus*).

Anthoptilum grandiflorum, a species that resembles *B. finmarchica*, is also reported at similar depths and latitudes. In the former, the autozooids on the lower part of the colony are grouped in pairs, not singly as in *B. finmarchica* (Opresko 1980). *Balticina finmarchica* is reported from the North Atlantic as far south as Cape Cod at depths between 120 and 2200 m. Our records fall within reported ranges of depth and latitude (Deichmann, 1936; Miner, 1950; Hecker et al., 1980).

Pennatula

This feather-shaped, red-coloured sea pen was recorded in video and photos from the Gully and Haldimand Canyon anchored in soft (muddy) bottom with occasional gravel and outcrops. It was moderately abundant in localized areas, but estimates may be confounded by its ability to withdraw completely into the sediment (Langton et al., 1990). Its depth of occurrence ranged between 210 and 441 m, and the only other co-occurring coral in the same general area was Alcyonacea-type 2. Commensals were not apparent in video or photos, and other epifauna was sparse, consisting of burrowing anemones and sporadic patches of large benthic foraminifera. Observed fish included *Urophycis chesteri*, *Sebastes*, flounder (Pleuronectidae) and an unidentified small fish that might be juvenile grenadier (Macrouridae).

Two species with overlapping distributions at these depths and latitudes are reported: *Pennatula aculeata* and *P. borealis*. *Pennatula borealis* is larger and colour is more variable, but others have found it impossible to distinguish between the two in photos and video (Hecker et al., 1980). This genus has been reported from Newfoundland to the Bahamas at depths between 120 and 2500 m, where it can be a major component of the benthic megafauna (Langton et al., 1990). Present records occur within the historical range of depth and latitude for this genus (Whiteaves, 1901; Deichmann, 1936; Miner, 1950; Hecker et al., 1980).

Undetermined white sea pen

This species was recorded in video and photo at the Stone Fence at a depth of 225 m. It was observed anchored in soft (muddy) bottom next to cobbles with Alcyonacea-type 2 and burrowing anemones. Commensals were not apparent in video or photos. This may be the same white-coloured sea pen mentioned by Opresko (1980), who described a delicate *Pennatula*-like sea pen at a depth between 150 and 300 m. The redfish, *Sebastes*, was observed.

The Scleractinians

Flabellum alabastrum Moseley, 1873

Flabellum alabastrum was the only scleractinian found during the surveys. It was recorded as incidental bycatch in groundfish trawls along the northeastern Scotian Shelf around Banquereau and Misaine Bank at depths between 231 and 237 m in areas reported to be sandy mud with gravel where it co-occurred with *Acanella arbuscula*. There were no commensals found on specimens of *F. alabastrum* examined. Invertebrate bycatch in groundfish trawls was insignificant, consisting of small amounts of mostly *Pandalus* and *Lithoides maja*. *Sebastes* and witch flounder (*Glyptocephalus cynoglossus*) dominated trawl biomass.

Flabellum alabastrum has been reported from Davis Strait to Georgia at depths between 357 and 1997 m (Cairns, 1981). These collections were shallow, but possibly not unique: *F. alabastrum* may also occur above 350 m in some canyons along the northeast coast of the United States (Hecker et al., 1980).

SUMMARY

The goal of this report was to summarize data collected on corals recently observed or collected by DFO in Atlantic Canada. It contains information on a total of 15 coral species and species groups from benthic surveys, observations at experimental sites, and from incidental bycatch in groundfish trawls (Table 1).

Corals were found at all areas surveyed and results are in general agreement with recent anecdotal observations (Breeze et al., 1997). The most abundant corals observed to date are on Romeys Peak, a relatively shallow area at the mouth of the Northeast Channel. This area has some of the strongest current velocities of any in the surveys. Observations in the channel were unique, with high densities of the coral *Primnoa resedaeformis* clearly related to slightly elevated areas of iceberg scour with abundant boulders and cobble. The highest diversities of corals were recorded at the Stone Fence (10 species) and the Gully (8 species), which appeared related to the wide variety of habitats surveyed at these locations. The northward extension of *Paragorgia arborea*, the collection of Gorgonacea-type 1, and the shallow distributions of *Acanella arbuscula*, *Acanthogorgia armata* and *Flabellum alabastrum* demonstrate the basic need for more knowledge on the distribution of coral.

New imaging gear developed at the Bedford Institute of Oceanography can now provide direct observations of corals in their natural habitat and information on associated species and community structure. These non-destructive observations are necessary for understanding the importance of corals as key or defining structures in their environment.

After reviewing video and photos, there did not appear to be disproportionate concentrations of fish near corals, or any unique fish species associated with corals. The adult fish fauna consisted of species that were expected to be common along the shelf and slope at similar depths (Scott and Scott,

1988). On no occasion were fish observed directly interacting with a coral colony (i.e., feeding off or sitting among branches). Juvenile (small) fish were either absent or else not apparent in and around corals as a result of low video resolution, the sampling technique or some behaviour. Patterns of habitat use and predator avoidance do suggest that tall biological habitat is not optimum for the survival of juvenile cod (Lindholm et al., 1999). Only the redfish (*Sebastes* sp.) appears to co-occur consistently with nearly all species of coral. This group of three species, *S. fasciatus*, *S. marinus* and *S. mentella*, occupies a wide range of habitats, occurring on rocky to clay bottoms. Redfish are pelagic or bathypelagic feeders, preying on pelagic crustaceans and smaller fish. They are known to swim off bottom at night, presumably to feed (Scott and Scott, 1988). Overall, there is no strong direct evidence to indicate that the densities and types of corals observed are locally important areas for feeding or as nurseries. Indirect information based on groundfish trawls suggests that fish assemblages may vary according to the species of corals present and whether colonies are dead or alive. More detailed analysis of benthic habitat is required to examine direct and indirect links between fish, habitat and coral.

Unlike fish, invertebrate species from six phyla were found interacting (commensal) on various corals. Most of the observed associations between coral and other fauna (commensal or otherwise) have yet to be fully understood, and many have probably yet to be observed. The ability of the videograb to selectively collect live, relatively undisturbed specimens and substrate may allow us to look more closely at commensal species and their particular relationship with the coral. Several coral colonies have been maintained alive and have already contributed to data on commensals and coral respiration rates. Tissue samples have been frozen for future work on genetics, and skeletal segments have been collected for various analyses.

Information on a wide variety of corals was collected, but peak abundances for many species are believed to be at depths greater than those surveyed to date. Though video and photos were well suited to surveying most Gorgonacea, many Alcyonacea and Pennatulacea could not be confidently identified. Concomitant sampling of the benthos and a detailed comparison with video and photos may be the only way to resolve the taxonomy of these groups. Study should proceed in the least intrusive manner possible, but a limited amount of precision sampling may greatly increase the quality and quantity of information.

These data are qualitative observations, not analyzed in their entirety, and incomplete in terms of the scope of surveys. It has been clearly demonstrated that we should continue to co-ordinate as many lines of information about corals as possible. We need to increase our current knowledge of distribution, abundance, habitat and associated species, as well as provide material for detailed taxonomic examination, local natural history collections, and public education. This report can be used to help better plan future analyses and studies of habitat and ecosystem dynamics.

ACKNOWLEDGEMENTS

We would like to thank the Photographic Unit at BIO for their invaluable technical support. Special thanks to Sheri Johnson, Kimberly Sheppard and Sylvia Rumbolt for help processing coral specimens, analyzing video and photos, archiving data, and suggestions during writing. We acknowledge the collection efforts of the technical staff in the Gadoids Section, Aquatic Resources Division at the Northwest Atlantic Fisheries Centre in Newfoundland, and the Marine Fish Division at BIO. We would like to thank the captains and crews of the vessels C.C.G.S. *Hudson* and *Parizeau* for their assistance, and all of the science staff who participated during the years of work at sea. We thank Shelly Armsworthy and Derek Fenton for reviewing this manuscript.

REFERENCES

Breeze, H., D.S. Davis, M. Butler and V. Kostylev. 1997. Distribution and status of deep-sea corals off Nova Scotia. Marine Issues Committee Special Publication Number 1. Ecology Action Centre, Halifax, N.S., 58 pp.

Cairns, S.D. 1981. Marine flora and fauna of the northeastern United States. Scleractinia. NOAA Technical Report NMFS Circular 438: U.S. Department of Commerce. National Oceanic Atmospheric Administration.

Cimberg, R.L., T. Gerrodette and K. Muzik. 1981. Habitat requirements and expected distribution of Alaska coral. Report prepared for Office of Marine Pollution Assessment, Alaska Office, Research Unit #601.

Deichmann, E. 1936. Alcyonaria of the western part of the Atlantic Ocean. Memoirs of the Museum of Comparative Zoology, Harvard University, 53 pp.

Energy, Mines and Resources Canada. 1986. Quaternary geology of the continental margin of eastern Canada. Geological Survey of Canada Map 1711A.

Fader, G.B., E. King, R. Gillespie, and L.H. King. 1988. Surficial geology of Georges Bank, Browns Bank, and the southeastern Gulf of Maine. Geological Survey of Canada Open File 1692, 3 sheets, scale 1:300,000.

Fader, G.B., D. J.W. Piper and C.L. Amos. 1998. Surficial, bedrock geology, and mapping. In G. Harrison and D. Fenton (eds), The Gully Science Review. DFO Canadian Stock Assessment Secretariat Research Document 98/83.

Greenberg, D.A. 1983. Modeling the mean barotropic circulation in the Bay of Fundy and Gulf of Maine, Journal of Physical Oceanography 13: 886-904.

Gregory, D.N., C. Bussard and S. Narayanan. 1996. Current statistics for the Grand Banks and Labrador Shelf. Canadian Data Report of Hydrography and Ocean Sciences 145.

Gregory, D.N. and C. Bussard. 1996. Current statistics for the Scotian Shelf and Slope. Canadian Data Report of Hydrography and Ocean Sciences 144.

Halliday, R.G. and P.A. Koeller. 1981. A history of Canadian groundfish trawling surveys and data usage in ICNAF Divisions 4TVWX. In W.G. Doubleday and D. Rivard. (eds), Bottom Trawl Surveys. Canadian Special Publications Fisheries and Aquatic Sciences 58, 273 pp.

Harrison, G. and D. Fenton (eds). 1998. The Gully Science Review, DFO Canadian Stock Assessment Secretariat Research Document 98/83.

Hecker, B., G. Blechschmidt and P. Gibson. 1980. Final report. Canyon assessment study in the Mid- and North Atlantic areas of the U.S. outer continental shelf. U.S. Department of Interior, Bureau of Land Management, Washington, D.C., Contract No. BLM AA551-CT8-49.

Kenchington, E.L.R, J. Prena, K.D. Gilkinson, D.C. Gordon Jr., K. MacIsaac, C. Bourbonnais, P. Schwinghamer, T.W. Rowell, D.L. McKeown and W.P. Vass. 2001. Effects of experimental otter trawling on the macrofauna of a sandy bottom ecosystem on the Grand Banks of Newfoundland, Canadian Journal of Fisheries and Aquatic Sciences 58: 1043-1057.

Langton, R.W., E.W. Langton, R.B. Theroux and J.R. Uzmann. 1990. Distribution, behaviour and abundance of sea pens, *Pennatula aculeata*, in the Gulf of Maine, Marine Biology 107: 463-469.

Lawrence, P., K.W. Strong, P. Pocklington, P.L. Stewart and G.B. Fader. 1985. A photographic atlas of the eastern Canadian continental shelf: Scotian Shelf, Grand Banks of Newfoundland. Maritime Testing (1985) Ltd., Dartmouth, N.S., viii, 187 pp.

Lindholm, J.B., P.J. Auster and L.S. Kaufman. 1999. Habitat-mediated survivorship of juvenile (0-year) Atlantic cod, Marine Ecology Progress Series 180: 247-255.

McKeown, D.L. and D.C. Gordon Jr. 1997. Grand Banks otter trawling impact experiment: II. Navigation procedures and results, Canadian Technical Report of Fisheries and Aquatic Sciences 2159: xi + 79p

Milligan, T.G., D. Belliveau, D.K. Muschenheim and G.R. Morton. 1998. New equipment for benthic habitat studies 1, Sea Technology, Sept 1998: 56-61.

Miner, R.W. 1950. Field book of seashore life. G.P. Putnam's Sons, New York.

Opresko, D.M. 1980. Taxonomic description of some deep sea octocorals of the Mid and North Atlantic, in B. Hecker, G. Blechschmidt and P. Gibson (eds), Final report. Canyon assessment study in the Mid- and North Atlantic areas of the U.S. outer continental shelf. U.S. Department of Interior, Bureau of Land Management, Washington, D.C., Contract No. BLM AA551-CT8-49.

Petrie, B., K. Drinkwater, D. Gregory, R. Pettipas and A. Sandstrom. 1996. Temperature and salinity atlas for the Scotian Shelf and the Gulf of Maine, Canadian Technical Report of Fisheries and Aquatic Sciences 171: v + 398 pp.

Petrie, B., J. Shore, C. Hannah and J. Loder. 1998. Oceanography. In G. Harrison and D. Fenton (eds), The Gully Science Review, DFO Canadian Stock Assessment Secretariat Research Document 98/83.

Prena, J., P. Schwinghamer, T.W. Rowell, D.C. Gordon Jr., K.D. Gilkinson, W.P. Vass and D.L. McKeown. 1999. Experimental otter trawling on a sandy bottom ecosystem of the Grand Banks of Newfoundland: analysis of trawl bycatch and effects on epifauna, Marine Ecology Progress Series 181: 107-124.

Rowell, T.W., P. Schwinghamer, M. Chin-Yee, K. Gilkinson, D.C. Gordon Jr., M. Hartgers, D.L. McKeown, J. Prena, D.P. Reimer, G. Sonnichsen, G. Steeves, W.P. Vass, R. Vine and P. Woo. 1997. Grand Banks otter trawling experiment: III. Sampling equipment, experimental design, and methodology, Canadian Technical Report of Fisheries and Aquatic Sciences 2190: viii + 36 pp.

Scott, W.B. and M.G. Scott. 1988. Atlantic Fishes of Canada, Canadian Bulletin of Fisheries and Aquatic Sciences 219, 731 pp.

Tendal, O.S. 1992. The North Atlantic distribution of the octocoral *Paragorgia arborea* (L., 1758) (Cnidaria, Anthozoa), Sarsia 77: 213-217.

Valentine, P.C. and R.G. Lough. 1991. The sea floor environment and the fishery of eastern Georges Bank: the influence of geologic and oceanographic environmental factors on the abundance and distribution of fisheries resources on the northeastern United States continental shelf. Open-file report 91- 439, U.S. Geological Survey, Department of the Interior.

Verrill, A.E. 1885. Results of the explorations made by the steamer "Albatross," off the northern coast of the United States, in 1883, in United States Commission of Fish and Fisheries, 1883 Commissioner's Report 11: 503-699.

Whiteaves, J.F. 1901. Catalogue of the Marine Invertebrata of Eastern Canada. Geological Survey of Canada, Ottawa.

OBSERVATIONS BY SUBMERSIBLE ON THE BIOCOENOSIS OF THE DEEP-SEA CORALS OFF PORTOFINO PROMONTORY (NORTHWESTERN MEDITERRANEAN SEA)

Leonardo Tunesi[1], Giovanni Diviacco[2], Giulia Mo[1]
[1] ICRAM, Rome, Italy
[2] Ufficio Parchi – Regione Liguria, Genova, Italy

SUMMARY
The Gulf of Tigullio is an area of high natural interest, and the biological richness of its benthic habitats is related to the strategic position of the Portofino Promontory in the northwestern Mediterranean. The benthic assemblages of the deep part of the area's circalittoral zone, and especially those found near the margin of the continental shelf-edge, are especially interesting. The importance of the hard-bottom, deep-water assemblages was first highlighted at the end of the 1950s, both for the presence of offshore rocky bottoms and for the deep-sea, white-coral banks. At that time the coral banks were described as having two facies, respectively characterized by *Madrepora oculata* (at a depth of 200 m) and by *Lophelia pertusa* (at 600 m). This biocoenosis was considered important from both a scientific and a fishing perspective.

In 1994 a study of the above-mentioned aspects was undertaken. Dives were carried out in three places off the Gulf of Tigullio with the aid of IFREMER's submersible *Cyana*. Surveys were conducted in the canyon at the mouth of the Entella River, specifically off Chiavari (at a depth of 570-100 m), off "Punta Chiappa" (at 140-280 m) and off "Punta di Portofino" (at 138-300 m).

The presence of live, deep-sea white corals in the area was confirmed, though the number of colonies was very small. There are two possible explanations for the infrequent occurrence of *M. oculata*: the area studied was rather small, and environmental factors such as temperature and sedimentation may have reduced the availability of surfaces suitable for this biocoenosis. The situation observed in the Gulf of Tigullio suggests the need for an integrated conservation approach to protecting these important habitats.

INTRODUCTION

The Gulf of Tigullio is an area of high natural interest, and the biological richness of its benthic habitat is related to the strategic position of the Portofino Promontory in the Ligurian Sea, northwestern Mediterranean (Tortonese, 1958; Rossi, 1969; Morri et al., 1986). Considerable logistical difficulties have hampered investigation of this area, so despite the scientific significance of the bathyal and the deep circalittoral hard-bottom assemblages, and of the biocoenosis of the detritic shelf-edge (DL) (Pérès and Picard, 1964) and the Entella River canyon, little is known about these habitats.

The first studies of deep-sea, hard-bottom assemblages in the area date back to the end of the 1950s (Rossi, 1958), and record the biocoenosis of the offshore rocky bottoms and deep-sea coral banks. Rossi (1969) obtained data on the presence of two *facies* of the deep-sea coral biocoenosis (CAP) (Pérès and Picard, 1964). The first, characterized by *Madrepora oculata*, occurs from the margin of the continental shelf into deeper water; the second, characterized by *Lophelia pertusa* (L.), is located at greater (600-700 m) depths. These clumps of deep-water scleractinian corals occur as branches that are either completely dead or else have only a few live terminal portions, and are considered relicts (Taviani and Colantoni, 1979). The structure and composition of these dead assemblages suggest that these scleractinian corals flourished when the deep waters of the Atlantic exerted a stronger influence on the Mediterranean, and the overall water temperature was colder, than is true today (Taviani and Colantoni, 1979). Delibrias and Taviani (1985) used ^{14}C analysis to demonstrate that the Mediterranean thanatocoenosis of these deep-water scleractinians dates back to the late Pleistocene (15,000-30,000 yrs B.P.), and hypothesize that the corals most likely died as result of reduced nutrient levels caused by an increase in temperature during the late glacial period. Clumps of live *M. oculata*, however, are not infrequent on canyon slopes in the northwest Mediterranean (Bourcier and Zibrowius, 1973); such areas may receive better nutrient input (Zabala et al., 1993).

This deep-sea coral biocoenosis is of great scientific importance from both a geological and a biological point of view. At present, however, there is very little information available concerning the vitality of the deep-water scleractinian corals and their associated benthic fauna, which in the Atlantic basin are otherwise very rich (Le Danois, 1948; Jensen and Rune, 1992). White-coral reefs are also of particular relevance to the fishing industry, whether it be in terms of the damage that they inflict on trawling gear (Wilson, 1979a), or of the highly correlated presence of species such as the European lobster (*Homarus gammaru*), the common spiny lobster (*Palinurus elephas*) and the wreckfish (*Polyprion americanus*), which generally prefer hard bottoms such as these.

In addition, there is no information on the biological assemblages in the canyon in front of the mouth of the Entella River, which reaches a maximum depth of 560 m, after which it merges with the deeper canyon of the Levant Riviera (Fanucci et al., 1979).

A study was therefore undertaken in 1994 to identify and evaluate the conservation status of the above-mentioned *facies*.

METHODS

In September 1994, with the aid of IFREMER's submersible *Cyana*, we made observations, collected biological samples, and took photographic and video images. Dives were carried out in three places off the Gulf of Tigullio (Figure 1). The study was part of the research program CYACHUM - LICYA (contribution EEC - DG XII - Program "Human Capital and Mobility"). The

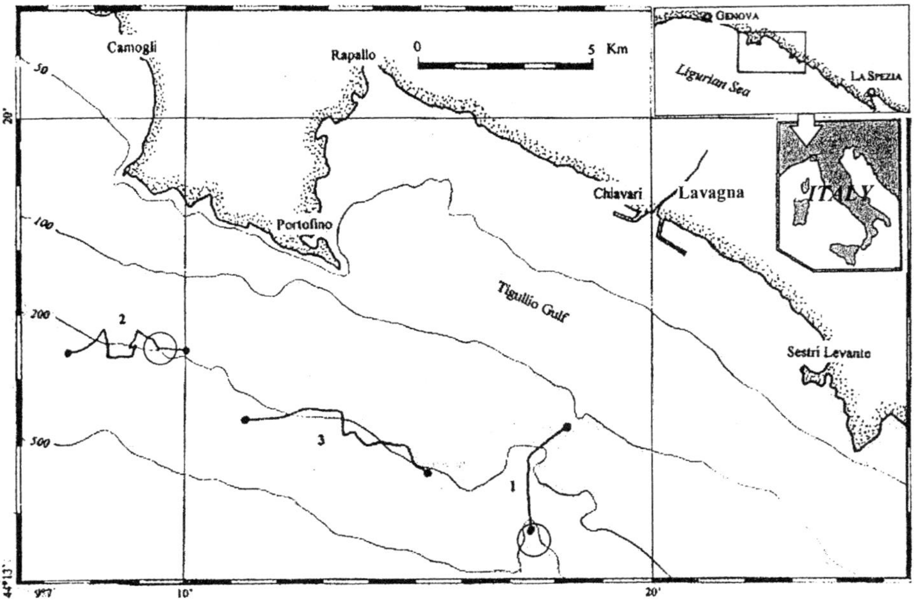

Figure 1. Location of the three dives conducted by the submersible Cyana *on bottoms seawards of the Gulf of Tigullio (1 = 12 September; 2 = 13 September; 3 = 14 September 1994). Circles zones indicate areas in which* Madrepora oculata *was found.*

exact position of the submersible during the dive was obtained by DGPS, with the application of two buoys positioned in the study area. During the dive, hard-bottom outcrops were searched with panoramic sonar having a 500 m range of action.

The investigations were carried out in the canyon off the mouth of the Entella River and specifically off Chiavari (at depths between 570 and 100m), off "Punta Chiappa" (at 140-280 m) and off "Punta di Portofino" (at 138-300m) (Figure 1 and Table 1). Dive sites were selected from information in the literature (Rossi, 1969) and by analysing fishing maps (International Seaway Marine cartography, 1988a,b). A ban by the Italian Ministry of Foreign Affairs made it impossible to carry out dives scheduled for deeper bottoms, where a thanatocoenosis of *Lophelia pertusa* is recorded. The visualized areas scanned during each dive were calculated on the basis of the visual width of 5 m provided by the submersible headlights.

The macrobenthic and nektonic species were later identified by analysing photographic and video material obtained in each dive. Benthic samples were selectively collected with the aid of *Cyana*'s pincer grab and later analysed in the laboratory. Organism densities were calculated using the length of the

Table 1. Information on dives conducted by submersible (DL = detritic shelf-edge)

Date	12.9.1994 (Dive 1)	13.9.1994 (Dive 2)	14.9.1994 (Dive 3)
Area	Entella Canyon	Bottoms off Punta Chiappa	Bottoms off Punta di Portofino
Explored range of depth	100 - 570 m	140 - 280 m	138 - 300 m
Length in hours	4	5	5
No. of identified taxa	21	16	55
Length of bottom investigated per dive	3.8 km	5.5 km	6.3 km
Total bottom surfaces investigated per dive	19000 m^2	275000 m^2	31500 m^2
Bottom surfaces hosting specific biocoenosis			
Deep-sea white corals	-	1 m^2	-
Offshore rocky bottoms	-	99 m^2	-
DL biocoenosis	-	9765 m^2	-
DL - Facies with *L. phalangium*	5700 m^2	7765 m^2	10175 m^2
DL - Conchiferous bottoms	-	2000 m^2	-
DL - Facies with *O. ophiura*	2850 m^2	-	6260 m^2
DL - Facies with *A. irregularis* penth.	-	-	2740 m^2
Bathyal mud	10450 m^2	17635 m^2	7045 m^2
Bathyal mud with *F. quadrangularis*	-	-	4305 m^2
Anthropogenic substrata	-	-	725 m^2

submersible's arm with respect to the observed field of view. Wrecks and remnants of objects derived from human activities were considered as hard bottoms of anthropogenic origin and were the subject of a separate species census.

RESULTS AND DISCUSSION

Dives in the Gulf of Tigullio (Table 1) allowed observations of different habitats and identification of 69 taxa belonging to Porifera (2), Cnidaria (8), Mollusca (10), Annelida (4), Crustacea Decapoda (8), Echinodermata (8), Chondricthyes (2) and Osteichthyes (27). Forty-nine taxa were found associated with soft bottoms, 24 with hard bottoms and 12 with anthropogenic substrata (Table 2).

Deep-sea White (Scleractinian) Corals

In the Mediterranean Sea *Madrepora oculata* is considered a bathyal species, and represents one *facies* of the deep-sea coral biocoenosis (CAP) (Pérès and Picard, 1964; Pérès, 1967). Live colonies of this species, however, have been found from the deep circalittoral zone (130 m in the Marseille area; Bourcier and Zibrowius, 1973) to depths as great as 1000 m (Zibrowius, 1980). The bathymetric range of this species in the Atlantic is even wider, ranging from 80 to 1500 m (Zibrowius, 1980).

Lophelia pertusa, a rather primitive species among modern corals (Shelton, 1980), represents another facies of the deep-sea coral biocoenosis (CAP). Live clumps of *L. pertusa* have been sampled in the Mediterranean in the area of Banyuls and in the Ionian Sea (Zibrowius, 1980), but in the Ligurian Sea, clumps of this species are probably composed only of dead, sub-fossil branches covered by Fe-Mn oxides (thanatocoenosis: Rossi, 1958). In deep waters of the Hebrides Sea, banks of this species may reach 18 m in height (Wilson, 1979a) and are ring-shaped (Wilson, 1979b).

Both *M. oculata* and *L. pertusa* are passive suspension-feeders, and occur only in places where the bottom is sufficiently steep to prevent mud sedimentation (Pérès, 1967).

In the present study *M. oculata* was found during dives conducted on September 12 and 13, 1994. On the first day, dead branches were found at the base of the canyon at a depth of 561 m. Terminal portions of these dead branches, which were lying on the sediment, were still alive. Small colonies of live *M. oculata* were observed during the second dive on natural hard substratum at a depth of 210 m. No colonies of *L. pertusa* were found during this cruise.

Offshore Rocky Bottoms

Rocky outcrops were found at a depth of 210 m during the dive conducted on September 13. The macrobenthic assemblage present on these outcrops was characterized by many specimens of the sea urchin *Echinus melo* and of a crustacean decapod belonging the genus *Munida*, but polyps of *Madrepora oculata* also occurred (Figure 1). The ichthyologic fauna was composed of species of commercial interest such as *Lophius piscatorius*, *Conger conger*, *Zeus faber* and *Helicolenus dactylopterus*, and by non-commercial species such as *Anthias anthias, Capros aper* and *Trisopterus minutus capelanus*.

Hard bottoms of anthropogenic origin (metallic buoys, tanks and a large unidentified structure) were observed in considerable quantities and recorded during the September 14 dive. The objects were disposed along a NW-SE axis, creating a 300-metre-long barrier. This habitat was characterized by the presence of an abundance of fish and lobsters (Table 2c), unlike the fauna occurring on soft bottoms also investigated during the *Cyana* dives (Table 2a). The biological assemblage of the former was characterized by species associated with hard-bottom assemblages and consisted of species of high commercial value, such as the European lobster (*Hommarus gammarus*), the common spiny lobster (*Palinurus elephas*) and the wreckfish (*Polyprion americanus*) (Table 2b).

Assemblages of the Soft Bottom: the Continental Shelf-edge and the Submarine Canyon of the Entella River

To increase the probability of encountering deep-sea white (scleractinian) corals, dives were conducted along the continental shelf-edge and in the Entella River canyon, locations where these corals are generally more frequent (Bourcier and Zibrowius, 1973). In reality, dense assemblages of the sea lily *Leptometra phalangium* characterized the continental shelf-edge seawards of some areas of the Gulf of Tigullio (Tunesi and Diviacco, 1997), reaching a density of 15-16 specimens per m^2, a density also recorded by Pérès (1958) for benthic habitats off France. The European wing oyster (*Pteria hirundo*) and the brittle pen shell (*Atrina fragilis*) were often associated with this species, confirming quantitative data on the macrobenthic species composition of this assemblage collected in the Gulf of Tigullio by means of trawl sampling (Peirano and Tunesi, 1986).

The dives also allowed us to identify and describe other habitats present in the transition between the circalittoral and the bathyal zone (Tunesi and Diviacco, 1997)

- two areas, quite limited in extent, characterized by sediments consisting of a mixture of sand, mud, and an important fraction of conchiferous sub-fossil debris belonging to a Quaternary thanatocoenosis
- an area with a great abundance of the asteroid *Astropecten irregularis pentacanthus*, a eurybathic species (Tortonese, 1972)
- one zone characterized by the presence of many colonies of the sea pen *Funiculina quadrangularis*, a species characteristic of the biocoenosis of the bathyal mud (VP) (Pérès and Picard, 1964) and indicative of the soft mud with superficial fluid film *facies* (Carpine, 1970). *F. quadrangularis* was accompanied by the brittle pen shell (*Atrina fragilis*), a species most abundant at depths of 180-240 m
- an area at depths of 157-211 m that coincided with the mouth of the

Table 2a. Species identified on soft bottoms during the dives, and the range of depth and relative abundance of each species (+ = presence; ++ = medium abundance; +++ = high abundance)

Species / Dive Date →	Range of depth			Relative abundance		
	12.9.94	13.9.94	14.9.94	12.9.94	13.9.94	14.9.94
SPERMATOPHYTA						
Posidonia oceanica leaves			185			+
PORIFERA						
Suberites domuncula Olivi			164			+
CNIDARIA			168			
Alcyonum palmatum (Pallas)	108-201			+		
Actinaria			156-205			++
Calliactis parasitica (J. Couch)			193			+
Cerianthus membranaceus Spallanzani	196-560		196-242	++		+
Epizoanthus arenaceus (Delle Chiaje)			202			++
Funiculina quadrangularis (Pallas)			173-257			++
Madreporaria			171			+
MOLLUSCA						
Atrina fragilis (Pennant)	119-247		120-198	++		+
Bivalvia	182			+		
Cassidaria sp.	261			+		
Eledone cirrhosa (Lamarck)	108	293	192-257	+	+	+
Glossus humanus (L.)			242			+
Loliginidae		269	192-299			+ +
Sepiola sp.			213			+
ANNELIDA						
Filograna sp.			198-200			++
Terebellides stroemi Sars			241			+
ARTHROPODA - Crustacea Decapoda						
Nephrops norvegicus (L.)	277-452			++		
Paguridae			193-203			++
Pasiphaea sivado (Risso)	480			+		
Plesionika sp.	533			+		
Portunidae			191			+
ECHINODERMATA						
Antedon mediterranea (Lamarck)			138-225			+
Astropecten irregularis pent. (D.Chiaje)			198-242			++
Leptometra phalangium (Muller)	108-233	284	120-243			+++
Ophiocten abyssicolum Mar.			200-225			++
Ophiura ophiura (L.)	157-211		182-257			+++
Spatangidae			179			+
Stichopus regalis Cuvier			194			++
VERTEBRATA - Chondrichthyes						
Centrophorus uyato (Rafinesque)	557			+		
Scyliorhinus canicula (L.)	227-474		224	+		
VERTEBRATA - Osteichthyes						
Arnoglossus laterna (Walbaum)			282			+
Buglossidium luteum (Risso)			193			+
Callyonimus maculatus Raf. Schmaltz			187-209			+
Cepola macrophthalma (L.)	112			+		
Citharus linguatula (L.)			225-246			+
Coelorinchus coelorincus (Risso)	441-545			+		
Gobidae			192-203			+
Gadiculus argenteus argenteus Guichenot			189-198			++
Gaidropsarus sp.			246			+
Lepidorhombus boscii (Risso)	202		171-395			++
Lesuerigobius sp.			188-243			+
Merluccius merluccius (L.)			159-206			+
Micromesistius poutassou (Risso)			186-316			+
Notolepis rossoi (Bonaparte)	562					+
Ophisurus serpens (L.)			255			+
Soleidae			201			+
Triglidae			149-213			+

Table 2b. **Species identified on hard bottoms during the dives, and the range of depth and relative abundance of each species (+ = presence; ++ = medium abundance; +++ = high abundance)**

Species		Range of depth			Relative abundance		
Species	Dive Date →	12.9.94	13.9.94	14.9.94	12.9.94	13.9.94	14.9.94
PORIFERA							
Porifera				166-201			+
CNIDARIA				168			+
Madrepora oculata L.		561	212-213		+	+	
Madreporaria				171			+
MOLLUSCA							
Eledone cirrhosa (Lamarck)		108	293	192-257	+	+	+
Octopus salutii Verany		258	214		+	+	
Ostrea sp.				180-184			+
Pteria hirundo (L.)		119-237		175-239	+		++
Loliginidae			269	192-299		+	+
ANNELIDA							
Sabellidae				164			++
Serpulidae			213	164-180		+	+
ARTHROPODA - Crustacea Decapoda							
Homarus gammarus (L.)				176-210			+
Munida sp.			212-214			++	
Palinurus elephas (Fabricius)				164-181			+
ECHINODERMATA							
Echinus melo Lamarck		108	213	189	++	++	++
Leptometra phalangium (Mull.)		108-233	284	120-243	++	+	++
VERTEBRATA - Osteichthyes							
Anthias anthias (L.)			213	171-213		+	++
Capros aper L.		229-238	214-252		++	++	
Conger conger (L.)			212-213	180		+	+
Helicolenus dactylopterus (Delaroche)		158-373	211-250	177-197	++	++	++
Lophius piscatorius L.			213			+	
Merluccius merluccius (L.)				159-206			+
Phycis phycis (L.)			213	164		+	+
Trisopterus minutus capelanus (Lacep.)			213	186-200		+	++
Zeus faber L.		108	213		+	+	

Entella River and the beginning of the canyon. Here *Leptometra phalangium* is replaced by an abundance of *Ophiura ophiura*. This situation seems to be due to the particular hydrodynamic conditions of the shelf-break area present at the mouth of the canyon.

Research on deep-sea white corals in the Entella canyon began on September 12. The submersible started from the base of the canyon, at a depth of 560 m. During this dive only two dead branches of *Madrepora oculata* were found at the base of the canyon, accompanied by a variety of debris of human origin.

The steep walls of the canyon are composed of a light grey mud, and did not exhibit a significant presence of benthic species. Specimens of the cylinder anemone (*Cerianthus membranaceus*) appeared along the canyon margins, while species characteristic of the upper part of the bathyal zone, such as the Norway lobster (*Nephrops norvegicus*), were found when the submersible moved away from the canyon margins toward the shelf-break edge. No hard

Table 2c. Species identified on anthropogenic substrata during the dives, and the range of depth and relative abundance of each species (+ = presence; ++ = medium abundance; +++ = high abundance)

Species Dive Date →	Range of depth 12.9.94	13.9.94	14.9.94	Relative abundance 12.9.94	13.9.94	14.9.94
MOLLUSCA						
Eledone cirrhosa (Lamarck)	108	293	192-257	+	+	+
Ostrea sp.			180-184			++
ANNELIDA						
Sabellidae			164			+
Serpulidae		213	164-180		++	++
ARTHROPODA – Crustacea Decapoda						
Homarus gammarus (L.)			176-210			++
Palinurus elephas (Fabricius)			164-181			++
ECHINODERMATA						
Leptometra phalangium (Mull.)	108-233	284	120-243	++	+	++
VERTEBRATA - Osteichthyes						
Anthias anthias (L.)		213	171-213		+	+++
Labrus bimaculatus L.			164			++
Phycis phycis (L.)		213	164		+	++
Polyprion americanus (Bloch & Schn.)			167-168			+
Trisopterus minutus capelanus (Lacep.)		213	186-200		+	++

bottoms or hard corals were found during this cruise along the steep walls of the canyon.

Leptometra was seen at depths of 108-200 m. *Ophiura ophiura* is present along the canyon axis and is particularly abundant between 160 and 150m. The bottom becomes quite flat at depths less than 150 m and shows evidence of trawl activities from 110 m to shallower depths.

The ichthyological fauna in the deeper part of the canyon appeared rather poor. The only species recorded were *Notolepis rissoi, Coelorinchus coelorincus, Scyliorhinus canicula*, and the little gulper shark (*Centrophorus uyato*). As the submersible ascended, specimens of the rockfish (*Helicocenus dactylopterus*) were seen quite frequently, being more abundant near rocky bottoms. The boarfish (*Capros aper*) was also present.

CONCLUSIONS

Use of a submersible in the studied area allowed us to record the presence of live deep-sea white corals, although the number of colonies was very small. There are two possible explanations for the infrequent occurrence of *Madrepora oculata*: the area studied was rather small, and environmental factors such as temperature and sedimentation may have reduced the availability of surfaces suitable for this biocoenosis. In particular, temperature plays an important role in the horizontal and vertical distribution of *M. oculata* and *Lophelia pertusa*; these two comparatively stenothermal species are limited in their vertical distribution by temperature values of 13°C for *L. pertusa* and 15°C for *M. oculata* (Zibrowius, 1980). The comparatively high temperature of the Mediterranean Sea thus limits the presence of *L. pertusa* to depths greater than 300 m.

These scleractinians, moreover, are preferentially distributed on topographic irregularities characterized by faster currents and a low sedimentation rate, such as seamount peaks or prominent steps on canyon slopes. Ahermatypic corals are negatively influenced by high deposition rates of fine materials (Sartori, 1980). Their occurrence in the study area may thus be negatively influenced by possible modifications/alterations to sedimentation rates caused by two new nearby harbours (Chiavari and Lavagna). These harbours have in fact heavily modified sediment dynamics along this stretch of coast over the last two decades (AA.VV., 1979). The extraction activities of inland slate quarries may add to the problem.

The facies with *Leptometra phalangium* seems to be the most important assemblage on the continental shelf-edge of the studied area, confirming the hypothesis that in the Ligurian Sea this assemblage is a characteristic of the shelf edge rather than of the detritic shelf-edge (DL) biocoenosis (Cattaneo, 1981). Dense assemblages of *L. phalangium* were observed in connection with the shelf-edge, even at shallower depths such as at the mouth of the Entella canyon (108 m).

The soft bottoms of the Gulf of Tigullio are characterized by an absence of steep slopes, rocks or wrecks, and have been heavily exploited by trawling activities, as indicated by furrows caused by the "doors" of this fishing gear. The attached fauna on this kind of bottom is in fact very poor, implying that trawling may significantly modify the composition of the macrobenthic assemblages, eliminating the more delicate species (e.g., *Funiculina quadrangularis*), which are therefore found only on stretches of soft bottom protected from trawling by the presence of wrecks.

The megabenthic and nektonic assemblages in the area of the Entella canyon, and in particular on its very steep walls, appeared relatively poor. This may be due to the impact of sliding mud on the benthic fauna.

Investigation by submersible of the deep bottoms of the Gulf of Tigullio confirms the significance of this area and underlines the need to protect deep-sea corals from a complex variety of anthropogenic pressures, including trawling and the increase in sedimentation rate caused by harbour construction and quarry activities. The situation observed in the Gulf of Tigullio suggests the need for an integrated conservation approach to protecting these important habitats.

ACKNOWLEDGEMENTS

We are grateful to the crew of the submersible *Cyana* for their assistance at sea during submersible operations.

REFERENCES

AA.VV. 1979. Atti del convegno nazionale per la difesa del litorale di Chiavari, Lavagna e Sestri Levante dall'erosione marina. Comune di Chiavari, Comune di Lavagna, Comune di Sestri Levante, C.N.R. Progetto Finalizzato Conservazione del suolo, Roma, 225 pp.

Bourcier, M. and H. Zibrowius. 1973 (1972). Les "boues rouges" déversées dans le canyon de la Cassidaigne (région de Marseille). Observations en soucoupe plongeante SP 350 (juin 1971) et résultats de dragages, Téthys 4: 811-841.

Carpine, C. 1970. Ecologie de l'étage bathyal dans la Méditerranée Occidentale, Mémoires de l'Institut océanographique, Monaco 2: 1-146.

Cattaneo, M. 1981. Contribution à l'étude du plateau continental de la Mer Ligure et haute tyrrhénienne. II. Echinodermata. Cahiers de Biologie Marine, 22: 11-24.

Delibrias, G. and M. Taviani. 1985. Dating the death of Mediterranean deep-sea scleractinian corals, Marine Geology 62: 175-180.

Fanucci, F., G. Fierro, M. Firpo, L. Mirabile and M. Piccazzo. 1979. La piattaforma continentale della Liguria appenninica. Atti Convegno Scientifico Nazionale Progetto Finalizzato Oceanografia e fondi marini, C.N.R. Roma 3: 1275-1289.

International Seaway Marine cartography. 1988a. Carta da pesca e dell'ambiente marino da Imperia a Portofino. N 8. Delfino Editrice srl, Marina di Pisa.

International Seaway Marine cartography. 1988b. Carta da pesca e dell'ambiente marino da Portofino a Marina di Pisa. N 9. Delfino Editrice srl, Marina di Pisa.

Jensen, A. and F. Rune. 1992. The fauna associated with the bank-forming deepwater coral *Lophelia pertusa* (Scleractinaria) on the Faroe shelf, Sarsia 77: 53-69.

Le Danois, E. 1948. Les profondeurs de la mer. Payot, Paris, 303 pp.

Morri, C., C.N. Bianchi., V. Damiani, A. Peirano, G. Romeo and L. Tunesi. 1986. L'ambiente marino tra Punta della Chiappa e Sestri Levante (Mar Ligure): profilo ecotipologico e proposta di carta bionomica, Bollettino dei Musei e degli Istituti biologici dell'Università di Genova 52(suppl): 213-231.

Peirano, A. and L. Tunesi. 1986. Aspetti della biocenosi dei fanghi profondi nel Mar Ligure Centro-orientale, Nova Thalassia 8 (suppl. 3): 647-648.

Pérès, J.M. 1958. Rémarques générales sur un ensemble de quinze plongés effectuées avec le bathyscaphe F.N.R.S. III (Rés. Sc. Camp. Bathyscaphe F.N.R.S. III, 1954-1957), Annals de l'Institut Océanographique 34(4).

Pérès, J.M. 1967. The Mediterranean benthos, Oceanogr. Mar. Biol. Ann. Rev. 5: 449-533.

Pérès, J.M. and J. Picard. 1964. Nouveau manuel de bionomie benthique de la mer Méditerranée, Rec. Travaux de la Station marine d'Endoume 47 (Bull. 31): 3-137.

Rossi, L. 1958. Contributo allo studio della fauna di profondità vivente presso la riviera ligure di levante. Annali del Museo Civico di Storia Naturale di Genova "G. Doria", Genova, Vol. II, N 92:1-13.

Rossi, L. 1969. Considerazioni zoogeografiche sul Bacino N.W. del Mediterraneo con particolare riguardo al Mar Ligure, Archivio Botanico e Biogeografico Italiano, Vol. XLV, 4a serie, 14 (4): 139-152.

Sartori, R. 1980. Factors affecting the distribution of ahermatypic corals on the Mediterranean seafloor: a probabilistic study, Deep-Sea Research 27A: 655.663.

Shelton, G.A.B. 1980. *Lophelia pertusa* (L.): electrical conduction and behaviour in a deep-water coral, Journal of the Marine Biological Association, U.K. 60: 517-528.

Taviani, M. and P. Colantoni . 1979. Thanatocoenosis würmiennes associées aux coraux blancs, Rapports et procès verbaux des réunion de la Commission internationale pour l'exploration scientifique de la Mer Médit. 25/26(4): 141-142.

Tortonese, E. 1958. Bionomia marina della regione costiera fra Punta Chiappa e Portofino (Riviera ligure di Levante), Archivio di Oceanografia e Limnologia 11: 169-210.

Tortonese, E. 1972. L'echinofauna del piano batiale nel Golfo di Genova. Doriana, Supplemento agli Annali del Museo Civico di Storia Naturale di Genova "G. Doria", Genova 5(204): 1-7.

Tunesi, L. and G. Diviacco. 1997. Observations by submersible on the bottoms off shore Portofino Promontory (Ligurian Sea). Atti XII Congresso A.I.O.L. (Isola di Vulcano, 18-21 Settembre 1996) 1: 61-74.

Wilson, J.B. 1979a. The distribution of the coral *Lophelia pertusa* (L.) [*L. prolifera* (Pallas)] in the North-East Atlantic, Journal of the Marine Biological Association, U.K., 59: 149-164.

Wilson, J.B. 1979b. "Patch" development of the deep-water coral *Lophelia pertusa* (L.) on Rockall bank, Journal of the Marine Biological Association of the United Kingdom, 59: 165-177.

Zabala, M., P. Maluquer and J.G. Harmelin. 1993. Epibiotic bryozoans on deep-water scleractinian corals from the Catalonia slope (Western Mediterranean, Spain, France), Scientia Marina 57(1): 65-78.

Zibrowius, H. 1980. Les Scléractinarires de la Méditerranée et de l'Atlantique nord-oriental, Mémoires de l'Institut océanographique, Monaco 11: 1-284.

GROWTH AND RECRUITMENT OF AN ALASKAN SHALLOW-WATER GORGONIAN

Robert Stone and Bruce Wing
National Marine Fisheries Service, Alaska Fisheries Science Center, Auke Bay Laboratory
Juneau, Alaska, USA

SUMMARY

Alaskan gorgonian corals are believed to provide important structural habitat for juvenile groundfish, especially species of *Sebastes*. More than 20 species of gorgonian corals have been described in Alaskan waters but little is known about their biology. This study was initiated in July 1999 to collect basic biological information about the life history of a shallow-water gorgonian species. Colonies of *Calcigorgia spiculifera* were tagged at two sites in Chatham Strait, southeast Alaska, using SCUBA. Colonies were tagged above the holdfast with plastic, cinch-type tags. Each colony was video-recorded while lightly compressed between a sheet of clear plexiglass and a rigid vinyl sheet marked with a 1-cm grid. In the laboratory, computer image-analysis tools were used to capture and calibrate images of each colony and to measure the length (excluding polyps) of the central branch and several lateral branches. These baseline images will be used to measure growth during annual SCUBA surveys. Growth data collected after one year indicate that growth rates are slower than those reported for other gorgonians worldwide and that intraspecific growth is highly variable.

INTRODUCTION

At least 20 species of gorgonian corals inhabit Alaskan waters. Specimens of all but one species have been incidentally entangled in fishing gear (e.g., hook and line, longlines, trawls, crab pots, and fish traps) and detached from the seafloor (Cimberg et al., 1981; B. Wing, unpublished data). Several species attain large size and provide essential habitat in the form of structure and refuge for species of demersal fish and invertebrates. The effects of coral habitat alteration on benthic communities are unknown but may be substantial due to the reported longevity and slow growth rates of cold-water corals (Heikoop et al., 1998). The North Pacific Fishery Management Council is currently considering measures to establish several marine protected areas where gorgonian corals are abundant.

Most Alaskan gorgonians inhabit depths which limit *in situ* observations and the collection of measurements and consequently little is known about their life history. In this study we examine growth and recruitment of one shallow-water Alaskan gorgonian.

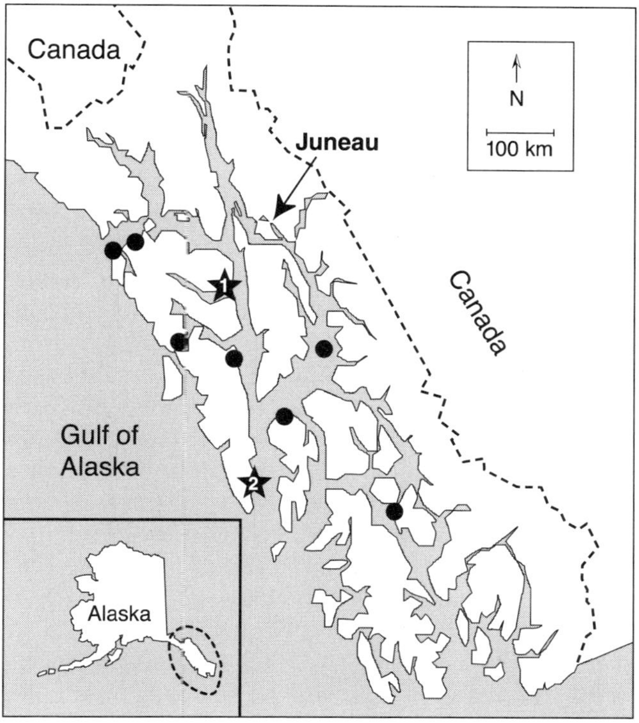

Figure 1. Map of southeast Alaska showing where growth of the gorgonian coral Calcigorgia spiculifera *was studied* in situ *at two sites (stars). This species of coral has been observed at seven other locations throughout southeast Alaska (circles).*

METHODS

We used SCUBA in July 1999 to tag colonies of *Calcigorgia spiculifera* Broch 1935. Colonies were tagged above the holdfast with numbered, plastic cinch-type tags. Twelve colonies were tagged at Site 1 and 23 colonies were tagged at Site 2 in 1999 (Figure 1). An additional three colonies were tagged at Site 1 in July 2000. Site 1 is near East Point, Chichagof Island, where colonies grow at a depth of 18-23 m. All colonies were attached to bedrock outcrops or boulders greater than 1 m diameter. Site 2 is 161 km south of Site 1 near Little Port Walter, Baranof Island (Figure 1). Most colonies at Site 2 were located on a bedrock slope and outcrop and a few were attached to boulders less than 1 m in diameter. Colonies at Site 2 grow at a depth of 23-29 m. Both sites had strong (>1 m sec^{-1}) tidal currents, and the rock substrate in the immediate area of the colonies had nearly 100% cover of bryozoans, hydroids and macroalgae. Coral colonies at both sites were restricted to an area of approximately 100 m^2.

Colonies were video-recorded in 1999 to provide baseline images from which annual growth increments could be determined. In 1999, colonies were video-recorded with a Sony TR-700 Hi-8[1] recorder. In 2000, colonies were video-recorded with a Sony TRV-900 digital recorder set in progressive scan mode. Colonies were video-recorded while gently compressed between a sheet of clear plexiglass and a rigid vinyl sheet marked with a 1-cm grid. *Calcigorgia spiculifera* has branches in a single plane, which allows the colony to be laterally compressed against the grid. The compression aligns the branches evenly against the measuring plane. Colonies relocated in 2000 were similarly video-recorded. In the laboratory, computer image analysis tools (OPTIMAS, 1999) were used to capture and calibrate images of each colony and to measure the length (excluding polyps) of the central branch and several lateral branches. Each branch was measured along the medial axis from the point opposite its origin. Branches were chosen for measurement based exclusively on whether there were clear images of the branches for both years. All branch lengths were computed as the average of three independent measurements made by the same person. Measurement error was computed as the positive discrepancies between the three measurements and the mean. We calculated a mean error of ±0.46 mm for all linear measurements made on computerized images in this study. Error increment appeared to be independent of branch length. We measured two components of growth: extension of existing branch tips and generation of new branches. Other measurements included height of the colony above the substratum and total width of the colony (uncompressed).

Water temperature was recorded every six hours with a temperature logger deployed at the maximum depth of coral at each study site. Four daily temperature measurements were averaged and presented as mean daily temperature. Salinity was measured during each sampling period with a Seabird Electronics Seacat Profiler. Diver observations included data on faunal associations, recruitment to artificial and natural substrates, predation and mortality. A 1-m^2 PVC frame fitted with a 1-m section each of stretched steel cable, rubber hose and polypropylene rope was placed at the base of Site 2 during July 1999 to serve as artificial settling substrata.

RESULTS AND DISCUSSION

Calcigorgia spiculifera is the most common shallow-water gorgonian in southeast Alaska. We have observed colonies of this gorgonian at nine locations throughout southeast Alaska at depths ranging from 18 to 306 m (Figure 1). Observations were made with SCUBA, a remotely operated vehicle, and the submersible DELTA. These sites are widely dispersed and colonies at each site inhabit a well-defined area. All sites consist predominantly of bedrock or large boulder substrate and have strong tidal currents.

Both study sites had similar seasonal patterns of mean water temperature between July 1999 and July 2000 (Figure 2). Temperature ranged from 4.1

[1] Use of trade names does not imply endorsement by the National Marine Fisheries Service.

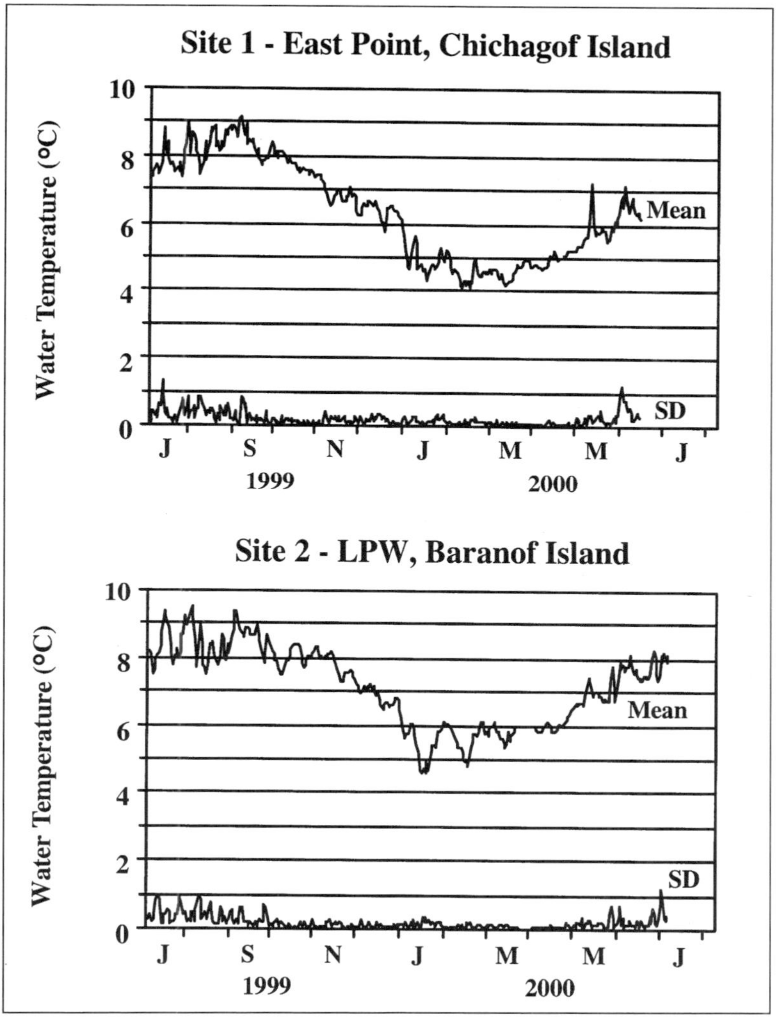

Figure 2. Mean daily water temperature at two sites where growth of the gorgonian coral Calcigorgia spiculifera *was studied. Temperature was recorded every 6 h with temperature loggers. SD is the standard deviation of the four daily measurements.*

to 10°C at Site 1 and from 4.4 to 9.8°C at Site 2. The mean annual temperature was 6.4°C and 7.1°C at Sites 1 and 2, respectively. Both sites showed little variation in daily water temperature (Figure 2). Salinity was 31.3 PSU at Site 1 and 31.8 PSU at Site 2 during both the 1999 and 2000 SCUBA surveys.

Baseline video footage was collected from 11 tagged colonies at Site 1 and 13 tagged colonies at Site 2 in July 1999. Video footage was collected from 13 colonies at Site 1 during July 2000, including nine for which we had 1999 baseline measurements. Video footage was collected from 18 colonies at Site 2 in July 2000, including 12 of those for which we had 1999 baseline measurements. We were able to measure growth from seven colonies at Site 1 and from nine colonies at Site 2 after one year. We were unable to make accurate growth measurements for the remaining colonies due to poor image quality.

Mean branch growth rate at both sites ranged from -1.8 to 14.8 mm yr^{-1} (Table 1). Individual branch growth rates for all colonies measured ranged from -4.3 to 19.2 mm yr^{-1} (Table 1). The small negative growth rates measured for some branches in this study may have been an artifact of our measurement technique, although negative growth rates have previously been reported for gorgonians (Yoshioka and Yoshioka, 1991). Growth rate was variable for branches from the same colony and also between colonies (Table 1). Highly variable intraspecific growth rates have been reported for other gorgonians (Yoshioka and Yoshioka, 1991). In general, growth rates measured during this study were much lower than those reported for other gorgonians, including the deep-water *Primnoa* sp. in Alaska (Andrews et al., 2000). Growth rates for *Primnoa* sp. were estimated at 17.4 mm yr^{-1} based on radio-

Table 1. Growth of *Calcigorgia spiculifera* measured between July 1999 and July 2000. Growth per colony was measured as the average growth of individual branches. Colony height is the length of the main (i.e., parent) colony branch.

Colony Tag No.	Colony Height (mm)	No. of Branches Measured	Mean Growth (mm)	Growth Range (mm)
Site 1 - East Point, Chichagof Island				
1411	132.6	5	0.3	(-2.9, 4.0)
1415	167.4	3	-1.8	(-2.3, -0.5)
1401	188.0	5	6.3	(2.5, 13.3)
1414	192.6	1	3.2	
1407	222.5	3	4.2	(1.3, 9.0)
1421	233.0	3	9.9	(6.3, 11.8)
1410	260.0	1	14.2	
Site 2 - LPW, Baranof Island				
1460	115.3	4	6.1	(-4.0, 14.4)
1462	116.0	2	0.6	(0.0, 1.1)
1461	132.4	4	-0.2	(-4.3, 3.8)
1483	149.4	4	5.3	(2.2, 9.5)
1481	151.3	3	14.8	(10.0, 19.2)
1472	159.9	4	9.6	(7.4, 11.5)
1453	167.0	4	6.5	(4.3, 8.0)
1498	183.0	5	3.4	(-1.5, 10.9)
1487	190.6	3	10.5	(0.4, 17.2)

metric and skeletal growth pattern analyses. Gorgonian studies in warmer waters have reported branch growth rates between 4.5 mm yr^{-1} and 62.1 mm yr^{-1} (Velmirov, 1975; Brazeau and Lasker, 1992).

In July 2000, two tagged colonies were missing at Site 1 and one tagged colony was missing at Site 2. These missing colonies may have become detached from the seafloor. Additionally, two tagged colonies were detached from the seafloor by SCUBA divers during the July 2000 survey. Yoshioka and Yoshioka (1991) reported that detachment was the major source of mortality for gorgonians at two Caribbean sites. They estimated a 27% detachment rate during their 5-year study. Although four colonies at Site 2 had areas of bare skeletal tissue or necrosis, no predation on colonies was observed.

Colonies smaller than 100 mm were not observed at either site in either year, indicating that recruitment had probably not occurred for at least several years (i.e., given a typical gorgonian growth rate of 20 mm yr^{-1}, a colony with a height of 100 mm would be five years old). The artificial substrata placed at Site 2 supported a lush growth of hydroids and bryozoans after one year, but no colonies of gorgonians were observed. These observations may indicate that recruitment in this species, at least at our two study sites, is a sporadic rare event.

We found that our measurements of colony height and width were ambiguous and were often dependent on the orientation of the colony between the plexiglass sheet and grid. Also, while small colonies (<180 mm) are typically uniplanar, larger colonies become increasingly arborescent and difficult to position uniformly against the measurement grid. Counts of individual branches were not possible for larger colonies, and we therefore excluded that measure of growth from analysis. Furthermore, identification of the main colony branch (i.e., parent branch) was often difficult. Our most reliable estimates of growth were for individual, lateral branches. Brazeau and Lasker (1992) similarly concluded that the best estimates of growth rate for colonies of the Caribbean gorgonian *Briareum asbestinum* were obtained by measuring individual branches.

Previous research on gorgonian growth measured *in situ* has been accomplished by making direct measurements of branch or colony length with calipers or rulers (Grigg, 1974; Velmirov, 1975; Yoshioka and Yoshioka, 1991; Brazeau and Laster, 1992). These measurements are typically made to the nearest 0.5 to 1.0 cm. We are unaware of other studies that have used calibrated video images to measure gorgonian growth *in situ*. This method has the advantage of providing a permanent record of colony morphometry and also greatly reduces diver time spent at depth. Highly accurate measurements are possible with proper colony orientation with respect to the calibration grid and parallel alignment of the grid with the camera lens. Accurate measurements of colony height for arborescent species, such as *Calcigorgia spiculifera*, may not be possible with this method, and growth measurements may be limited to individual branches for larger colonies. Additionally, accurate linear measurements are difficult to obtain from colonies in which the polyps are extended while feeding. This was a problem for *Calcigorgia spiculifera*, which has relatively large polyps (approximately 1 mm in length).

The slow growth rates measured during the first year of this study, although preliminary, are noteworthy since shallow-water corals are widely believed to have faster growth rates and shorter life spans than deep-water corals. Additionally, recruitment appears to be a rare, sporadic event. Shallow-water gorgonian communities may therefore exhibit slow recovery rates from seafloor perturbations. Our future research priority is to establish a third study site where colonies are more numerous and more variable in size (and therefore age).

ACKNOWLEDGEMENTS

We thank Ken Krieger, Linc Freese, Alex Wertheimer, Chuck O'Clair and Chris Lunsford for their assistance in the field and Jeff Regelin for performing the computer image analyses. This project was funded by the Auke Bay Laboratory, Alaska Fisheries Science Center, National Marine Fisheries Service.

REFERENCES

Andrews, A.H., E. Cordes, M.M. Mahoney, K. Munk, G.M. Cailliet, K.H. Coale, J. Heifetz and D. Somerton. Age and growth of a deep-sea, habitat-forming octocorallian (*Primnoa* sp.) from the Gulf of Alaska, with radiometric age validation, Hydrobiologia (in press).

Brazeau, D.A. and H.R. Lasker. 1992. Growth rates and growth strategy in a clonal marine invertebrate, the Caribbean octocoral *Briareum asbestinum*, Biological Bulletin 183: 269-277.

Cimburg, R.L., T. Gerrodette and K. Muzik. 1981. Habitat requirements and expected distribution of Alaska coral. Final Report Research Unit #601 to Office of Pollution Assessment, Alaska Office. VTN Oregon, Inc., 54 pp.

Grigg, R.W. 1974. Growth rings: Annual periodicity in two gorgonian corals, Ecology 55: 876-881.

Heikoop, J.M, M.J. Risk, A.V. Lazier and H.P. Schwarcz. 1998. ∂^{18}O and ∂^{13}C signatures of a deep-sea gorgonian coral from the Atlantic coast of Canada, Spring American Geophysical Union, Boston Meeting, Abstract.

OPTIMAS. 1999. Optimas 6.5 User Guide, Ninth Edition. Media Cybernetics, Silver Spring, Maryland, 548 pp.

Velimirov, B. 1975. Growth and age determination in the sea fan *Eunicella cavolinii*, Oecologia 19: 259-272.

Yoshioka, P.M. and B.B. Yoshioka. 1991. A comparison of the survivorship and growth of shallow-water gorgonian species of Puerto Rico, Marine Ecology Progress Series 69: 253-260.

THE ATLANTIC CORAL ECOSYSTEM STUDY (ACES): FORGING A NEW PARTNERSHIP BETWEEN SCIENTISTS AND PRINCIPAL STAKEHOLDERS

Anthony J. Grehan[1], André Freiwald[2] and the ACES Consortium.
[1] Martin Ryan Institute, NUI, Galway, Ireland
[2] Institut für Geologie und Paläontologie, Universität Tübingen, Germany

SUMMARY

The Atlantic Coral Ecosystem Study (ACES) is a multinational, interdisciplinary project funded for three years by the EU Fifth Framework Research Programme. It adopts a whole-ecosystem approach to the study of the poorly known but widely distributed European deep-water cold coral province. While recent research has clearly shown that *Lophelia* bioherms are extremely rich with respect to diversity, quantity and complexity of life forms, it is clear that great differences occur between geographic locations and that the structure and functioning of these systems are still poorly understood, locally as well as over the geographical range. The pan-European distribution of deep-water coral ecosystems found along the continental slope, deep shelf and fjord settings provides an ideal opportunity to study these differences.

The EU Fifth Framework Programme, in an effort to increase the socio-economic impact of its research and development projects, strongly encourages the formation of scientist-stakeholder partnerships and development of a suitable research-product delivery mechanism. The ACES project will respond to these challenges in a number of innovative ways: (i) through the production of an end-user-friendly 'coral reef manual' on DVD-ROM containing a set of recommendations for sustainable use of the coral ecosystem, (ii) by forming an ACES project-stakeholder partnership principally through links established at ACES Stakeholder Consultative Workshops, and finally (iii) by improving public awareness of the cold coral ecosystem through provision of an information pack (with emphasis on educational material) in multimedia format, public access to the ACES Web site, through mass media reportage and displays of captive coral in public aquaria.

At the first ACES Consultative Workshop held on 23 June 2000 in Galway, Ireland, a number of important issues were raised. The use of destructive trawling techniques first seen in Norway is now alleged to occur in the deeper coral found off the Irish west coast. It was agreed that urgent action needed to be taken to quantify the extent of this practice and to implement conservation measures if necessary. In the face of considerable uncertainty as to the applicability of the EU Habitats Directive as enacted into Irish environmental law, it was decided to set up a national task force consisting of ACES members and representatives of the relevant Irish government departments to rapidly address this issue.

INTRODUCTION

Although deep-water corals have been known to occur along the European margin since the last century (Le Danois, 1948), their extent (Wilson, 1979; Zibrowius, 1980; Frederiksen et al., 1992; Rogers, 1999) and potential importance as a key structural element in the European deep-water biotope has only recently become apparent (Jensen and Frederiksen, 1992; Freiwald et al., 1997; Freiwald, 1998; Rogers, 1999). Advances in side-scan and multi-beam mapping technology combined with improved *in situ* exploration capabilities (principally ROVs and other imaging platforms) have revealed a hitherto unexpected realm of coral colonies, reefs and giant bioherms (Mortensen et al., 1995; Hovland et al., 1998; Freiwald, 1998; Henriet et al., 1998).

The steady downslope movement and subsequent increased depth of operation of Europe's hydrocarbon exploration and deep-sea fishing industries has for the first time brought these activities into the realm of deep-water coral. Concerns about the potential environmental damage to corals during oil and gas exploration (Rogers, 1999) and documented evidence of large-scale damage to reefs off Norway due to destructive fishing practices (Fosså et al., 2000), have brought sharply into focus the urgent need to develop a sustainable management strategy to protect the coral ecosystem.

The ACES project will provide a margin-wide environmental baseline assessment of the status of Europe's deep-water coral ecosystem and provide recommendations for essential monitoring and methodology requirements to support future sustainable development. The evolution of new management concepts for a sustainable use of deeper-water marine ecosystems on a margin-wide scale is a grand challenge that can only be achieved on a joint European scale (Annex V of the OSPAR Convention).

The ACES project is one of six Fifth Framework-funded research and development projects forming the Ocean Margin Research Consortium (OMARC). This cluster will increase the value of the individual projects through support of a focused research strategy along Europe's continental margin.

ACES CONSORTIUM

The ACES project numbers researchers from five European countries (Table 1), forming a multidisciplinary group of geologists, physical oceanographers, sedimentologists, biogeochemists, molecular geneticists and ecologists.

PRINCIPAL STUDY LOCATIONS

In the northeast Atlantic, the geographic distribution of deeper-water coral (hereafter DWC) ecosystems can be traced from the continental slopes and banks off the Iberian Peninsula (Le Danois, 1948) to as far north as the Norwegian Shelf (Dons, 1944; Hovland et al., 1994) (Figure 1). Under certain oceanic, topography-guided circulation patterns, these DWC systems also occur in shallow-water coastal regions along the Swedish Bohuslän coast (Lundalv, pers. comm.) and in Norwegian fjords (Strømgren, 1971).

To cover the variety of environmental factors and interactions at ocean boundaries which enable the development of DWC ecosystems, the ACES community will focus on selected key study locations along this latitudinal gradient: Galicia Bank, Porcupine Slope, Rockall Trough, Skagerrak and Norwegian Shelf (Figure 1, Table 2). These study sites will be investigated with standardized methods to achieve a high degree of comparability – a major prerequisite for providing recommendations for the sustainable use of this unique ecosystem.

SCIENTIFIC OBJECTIVES

Objective 1: To map the structural and genetic variability, the framework-constructing potential, and the longevity of DWC ecosystems

High-resolution maps of DWC buildups are essential to determine the spatial distribution and the status of the ecosystem in the various working areas. The framework-constructing potential of DWC largely depends on the annual

Table 1. The ACES Consortium.

Affiliation	Country
University of Tübingen	Germany (Co-ordinating Institute)
University of Ghent	Belgium
University of Liverpool	United Kingdom
Scottish Association for Marine Science	United Kingdom
Martin Ryan Institute, NUI, Galway	Ireland
University of Southampton	United Kingdom
Southampton Oceanography Centre	United Kingdom
Coastal Resource Centre, UCC	Ireland
The Netherlands Institute for Sea Research	Holland
Tjarno Marine Biological Laboratory, University of Gothenburg	Sweden
EcoServe Ltd. Dublin	Ireland
Marine Informatics, Dublin	Ireland
University of Erlangen-Nuernberg	Germany
Institut für Umweltphysik, Heidleberg	Germany

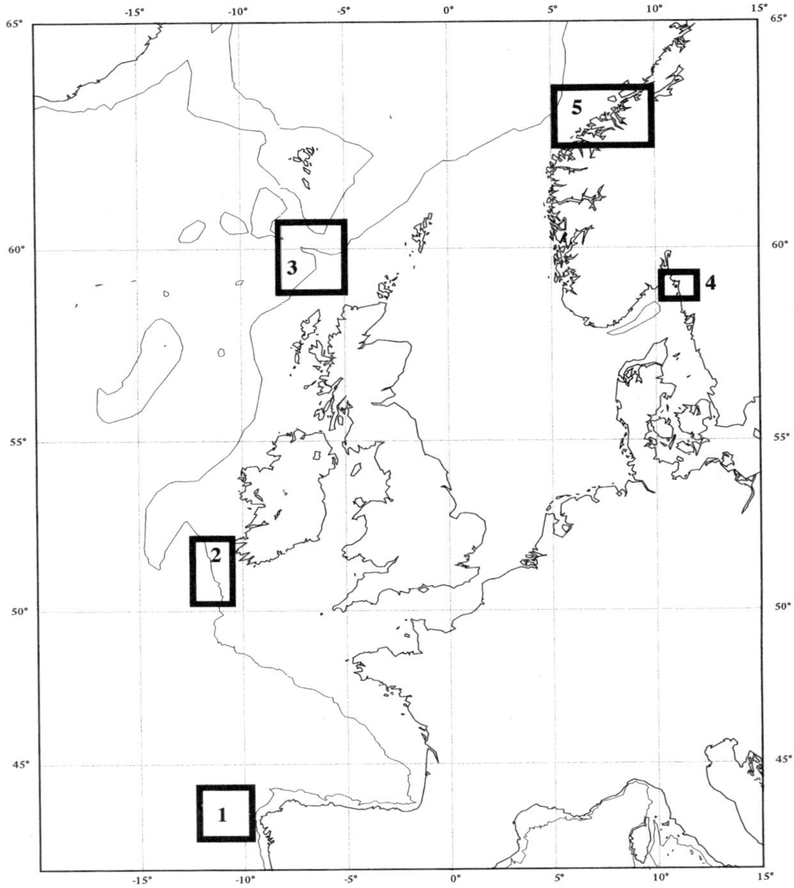

Figure 1. The five key ACES study locations found along the European Margin.

extensional growth rate, the intensity of secondary thickening of the coral skeleton and the intensity of post-mortem destruction by endolithic borers. We will utilize molecular genetic techniques to assess the spatial genetic structure and population dynamics of *Lophelia* at several scales (between regions, within regions and within individual coral reefs).

Expected Results: The outcome will be the first digitized atlas based on acoustic and video-imaging of different DWC buildup types along Europe's margin. Detailed growth rate, dating, and ultrastructural studies performed on *Lophelia* and *Madrepora* skeletons will provide results on the longevity of coral frameworks for each of the DWC working areas. The genetic techniques will enable us to identify levels of genetically effective migration between different geographic areas. It will also allow us to identify which *Lophelia* reefs contain the greatest genetic diversity. Both the larger scale imaging data and the detailed studies of the corals themselves will be integrated into a status check of the structural integrity of the DWC ecosystems.

Objective 2: To assess hydrographic and other local physical forcing factors affecting benthic boundary layer (BBL) sediment particle dynamics and particulate organic carbon (POC) supply in the vicinity of DWC ecosystems

DWC ecosystems are often found at or near oceanographic boundaries (even in fjords) but the detailed effect of hydrographic conditions on DWC buildups remains a matter of speculation. The poleward-flowing warm and saline NE Atlantic slope current is a well-documented feature at the shelf break that extends from the Iberian to the Norwegian Sea margin. Predominantly poleward (northward) slope currents at an eastern boundary tend to drive downward near seabed currents in the frictional layer. This has implications for the transport of suspended material in the benthic boundary layer (BBL) and hence for the nutrition and distribution of corals. In addition, the suspected contribution of hydrocarbon enrichment in the vicinity of the coral ecosystem will be assessed.

Expected Results: The goal is to provide a hydrographic description for the NE Atlantic margin where corals occur, including information on local hydrodynamics (currents) and water mass properties (temperature, salinity, density) at or near DWC ecosystems. A detailed description of suspended particle dynamics in relation to BBL flow and its impact on coral nutrition and feeding (either derived from *advected particulates* or *hydrocarbon venting*) is key to understanding the function of the whole DWC ecosystem. This will be achieved by molecular analysis of biologically labile particulate organic matter (POM) fluxes within the coral systems and surrounding surface sediments.

Objective 3: To describe the DWC ecosystem, its dynamics and functioning, investigate coral biology and behaviour, and assess coral sensitivity to natural and anthropogenic stressors

This objective takes a whole-ecosystem approach, addressing not only important aspects of coral biology such as reproduction, recruitment and feeding behaviour, but also intraspecific biotic interactions such as the importance of coral stands as refugia (particularly for juvenile commercially important fish species) in promoting the high associated biodiversity of the coral ecosystem fauna. Detailed food web analysis will help elucidate individual species' response to local variations in physical forcing and BBL organic carbon characteristics related to the presence of the coral framework. Coral sensitivity to natural and anthropogenic stressors will be determined both *in situ* and in controlled laboratory experiments.

Expected Results: The project should provide a comprehensive understanding of important aspects of coral biology and elucidate the role of the coral framework in promoting increased biodiversity when compared with the adjacent seabed. It should also provide a description of coral ecosystem food web relationships and resource partitioning related to variations in local physical forcing and BBL organic carbon inputs. Identification of coral health indicators in the field, and a development of experimental systems and protocols for toxicological and physical stress testing in the laboratory.

MANAGEMENT OBJECTIVES

Objective 1: Identification of the principal conservation concerns and management issues facing sustainable use of the DWC ecosystem and increasing public and political awareness of the urgent need to implement measures to protect and conserve the DWC ecosystem

Identification of the principal conservation concerns and management issues facing sustainable use of the DWC ecosystem will be achieved through consultation with the principal stakeholders following the establishment of an ACES-Stakeholder Partnership consisting of:

- representatives of the national bodies responsible for conservation, and fisheries and mineral/aggregate resource management along the European margin where the DWC ecosystem occurs
- representatives of the fishing and mineral/aggregate extraction industries
- the EU 5FP Research DG (through regular contractual reporting) and the European Environmental Agency (EEA), and
- appropriate non-government environmental organizations (e.g., WWF, IUCN, IOC) and intergovernmental organizations (ICES, OSPAR).

Two consultative workshops will facilitate Partnership exchanges. The first,

Table 2. Site characteristics of ACES targets along Europe's NE Atlantic Margin.

	Site				
	Galacia Bank	**Porcupine Slope**	**Rockall Trough**	**Kosterfjord (Sweden)**	**Sula Ridge (Norway)**
Latitude	43°N	51°N	59°N	59°N	64°N
Water depth (m)	700-900	600-1200	1000	85-150	315-248
Seabed structure	Coral thickets on giant pelagic ooze sand waves	Giant thickets on top of large carbonate mounds (up to 300 m high)	Complete coral cover of ca. 5-m-high mounds	Shallowest DWC occurrence ever reported. End member character	Largest known DWC reef in the NE Atlantic extending to 13 km

at the start of the project, has already been held (see below). Its purpose was to brief stakeholders on the science objectives, establish the partnership network, provide a forum for the exploration of opportunities for data exchange (particularly data not held in the public domain), examine potential links to other programmes, and highlight ACES research priorities through identification of the principal conservation concerns and management issues as identified by stakeholders and scientists. By holding the workshop at the very outset of the project, project management had the opportunity to adapt/respond to additional requirements.

The second workshop will take place at the end of the project, when a summary of research findings will be presented together with an assessment of the sensitivity of DWC ecosystem to perturbation (see below). Recommendations supporting sustainable environmental management of the DWC ecosystem will be made and the most practical form for the publication of the 'coral reef manual' (see below) will be discussed with end-users.

Increased public awareness of the environmental issues facing the DWC ecosystem will be achieved by permitting public access to parts of the ACES Web site; producing an educational DWC ecosystem DVD-ROM 'information pack' in multi-media format featuring video, photographs and graphics; through regular media reportage; and through captive display of coral specimens in public aquaria.

Expected Results: A new scientist-stakeholder partnership will be formed including legislators, scientists, environmental and resource managers, and representatives of offshore industries. Public and political awareness of the need to manage the DWC ecosystem will be increased.

Objective 2: Make recommendations for the sustainable use of the DWC ecosystem

This objective will be achieved in two steps:

i) Following synthesis of the research results gathered under the scientific objectives, the ability of the coral ecosystem to cope with various types of environmental impact will be assessed through the application of a species and ecosystem management information technology tool, i.e., the Marine Life Information Network for Britain and Ireland (MarLIN). MarLIN is a Web-based information technology tool currently under development in support of environmental management, protection and education (Hiscock et al., 1999). It has evolved over a number of years and is well supported. It provides a structure for linking available data on marine life around Britain and Ireland as well as providing a rigorous protocol for assessing species and ecosystem sensitivity to natural and anthropogenic perturbations. ACES will add key information to the database on the biology (e.g., taxonomy, general biology, distribution and habitat preference, reproductive biology and species importance) and sensitivity of priority DWC species. The information recorded will then be used to allocate scores of sensitivity and recovery potential for each selected species following assessment of their likely response to a range of specific natural and man-made impacts. More information about the MarLIN database and protocol can be found at http://www.marlin.ac.uk.

ii) Publication of an end-user-friendly 'deep-water coral reef manual' on DVD-ROM containing recommendations for the sustainable use of the coral ecosystem. A practical manual will be assembled containing recomendations, 'best practice' methodologies, and baseline data to be used as a tool for the sustainable management of the coral ecosystem. Feedback on end-user requirements from the Stakeholder Partnership will influence the range of information included and the presentation format.

Achievements: This will be the first sensitivity coding of a deep-water ecosystem based on impartial scientific evidence. It will produce guidelines and practical recommendations in the form of a deep-water 'coral reef manual' for future environmental monitoring and the sustainable use of the deep-water coral ecosystem. Production of the DVD-ROM management tool following consultation with the end-users will increase its practicality and applicability.

Recommendations for sustainable use of the DWC ecosystem will act as a primer for extending existing EU policy regarding coastal zone management to the deep-sea province.

FIRST ACES SCIENTIST-STAKEHOLDER FORUM

Given the strategic importance of Irish coral locations (including carbonate mounds), the first consultative workshop with principal stakeholders was held in Galway, Ireland, on 23 June 2000. The workshop was attended by representatives of all the major Irish off-shore stakeholder groups including representatives from the oil and gas industry, fishermen, tourism (public aquarium), national resource and environmental managers, and the national newspaper's marine correspondent. The workshop commenced with a presentation of the ACES project to the stakeholders. This was followed by presentations on related projects; invited speakers informed the forum of recent environmental studies supported by oil company consortia, i.e., the Irish Petroleum Infrastructure Programme (PIP) (John O'Reilly, Ireland) and the UK Atlantic Frontier Environmental Network (AFEN) (Gill Bishop, Scotland). Murray Roberts, Scotland, summarized the UK NERC-funded, Managing Impacts in the Marine Environment (MIME) project. An overview of the deep-water fisheries in the northeastern Atlantic (John Gordon, Scotland), and the effects of fishing activity on shallow *Lophelia* reefs in Norway and recent conservation legislation introduced there to reduce damage to reefs (Paal Mortensen, Norway), were also presented. The final session involved discussion of the current status of Irish legislation governing deep-water habitat conservation (Liz Sides, Ireland) and an overview of the difficulties posed for the creation of marine protected areas and legal protection of *Lophelia* within the existing international legal framework (Sabine Christiansen, WWF). Open discussions took place at regular intervals during the course of the presentations.

Some of the major outcomes of the meeting:

i) Agreement by all concerned that the DWC ecosystem must be conserved and an effective management strategy put in place.

ii) Support for the ACES project by oil company consortia with an undertaking to allow access to non-public-domain Tobi (side-scan) data if required.

iii) The major impact of bottom trawling was seen to be compounded by the adoption of destructive fishing techniques, i.e., 'rolling' of reefs prior to trawling. This practice already known to occur in Norway is alleged to now also take place in the deeper coral found off the Irish west coast (Paddy O'Malley [local fisherman] pers. comm.). It was agreed that there is a need to quantify the extent of trawl damage to reefs.

iv) Technical fishing conservation measures in the form of a trawling ban in favour of longlining (as in Norway) was mooted to protect Irish coral. However, it was pointed out that switching to longlining may lead to other conservation problems (e.g., over-fishing of sharks). There was some consensus that corals are a good candidate for conservation through designation of marine protected areas.

v) The urgent need to implement conservation measures in the Irish sector was highlighted. Discussions about the legal context in which this could be done raised a number of issues:

   a) Does the enactment of the EU Habitats Directive into Irish law give jurisdiction to the outer boundary of the 200-mile Exclusive Fisheries Zone while ratification of the Irish EEZ is pending?
   b) If the Irish legal instrument does provide for protection of the DWC coral ecosystem, which is the competent Irish government department with responsibility for implementation of conservation?
   c) What resources will be made available to enforce conservation measures.
   d) It was decided to set up a national task force consisting of ACES members and representatives of the relevant Irish government departments to rapidly address these issues.

CONCLUSIONS

The ACES project will not only address the sustainable environmental management of the DWC ecosystem. It will also act as a primer for the general development of an integrated off-shore environmental management strategy by stimulating dialogue between industry, conservationists and government and by focusing attention on the need to clarify the legal basis for implementation of conservation strategies.

ACKNOWLEDGEMENTS

The Atlantic Coral Ecosystem Study (ACES) is funded by the EU Fifth Framework Programme Contract No. EVK3-CT-1999-00008.

REFERENCES

Dons, C. 1944. Norges korallrev, Det Kongelige Norske Videnskabers Selskabs Forhandlinger 16: 37-82.

Fosså, J.H., P.B. Mortensen and D.M. Furevik. 2000. *Lophelia*-coral reefs in Norway. Distribution and effects of fishing, Fisken og havet No. 2, 94 pp.

Frederiksen, R., A. Jensen and H. Westerberg. 1992. The distribution of the scleractinian coral *Lophelia pertusa* around the Faroe Islands and the relation to internal tidal mixing, Sarsia 77: 157-171.

Freiwald, A. 1998. Geobiology of *Lophelia pertusa* (Scleractinia) reefs in the North Atlantic. Habilitation thesis, University of Bremen, 116 pp.

Freiwald, A., R. Henrich and J. Pätzold. 1997. Anatomy of a deep-water coral reef mound from Stjernsund, West-Finnmark, northern Norway, SEPM Special Publication 56: 141-161.

Henriet, J.P., B. De Mol, S. Pillen, M. Vanneste, D. Van Rooij, W. Versteeg, P.F. Crocker, P.M. Shannon, V. Unnithan, S. Bouriak and P. Chachkine. 1998. Gas hydrate crystals may help build reefs, Nature 391: 64-649.

Hiscock, K., A. Jackson and D. Lear. 1999. Assessing seabed species and ecosystems sensitivities. Existing approaches and development. Report to the Department of the Environment Transport and the Regions from the Marine Life Information Network (MarLIN), Marine Biological Association of the UK, Plymouth (MarLIN Report No.1), October 1999 edition.

Hovland, M., R. Farestveit and P.B. Mortensen. 1994. Large cold-water coral reefs off mid-Norway: a problem for pipe-laying?, in: Conference Proceedings (3), Oceanology International, 8-11 March 1994, Brighton.

Hovland, M., P. B. Mortensen, T. Brattegard, P. Strass and K. Rokoengen. 1998. Ahermatypic coral banks off mid-Norway: evidence for a link with seepage of light hydrocarbons, Palaios 13: 189-200.

Jensen, A and R. Frederiksen. 1992. The fauna associated with the bank-forming deep-water coral *Lophelia pertusa* (Scleractinia) on the Faroe shelf, Sarsia 77: 53-69.

Le Danois, E. 1948. Les profondeurs de la mer. Payot, Paris, 303 pp.

Mortensen, P.B., M. Hovland, T. Brattegard and R. Farestveit. 1995. Deep-water bioherms of the scleractinian coral *Lophelia pertusa* (L.) at 64°N on the Norwegian shelf: structure and associated megafauna, Sarsia 80: 145-158.

Rogers, A.D. 1999. The biology of *Lophelia pertusa* (Linnaeus 1758) and other deep-water reef-forming corals and impacts from human activities, International Review of Hydrobiology 84: 315-406.

Strømgren, T. 1971. Vertical and horizontal distribution of *Lophelia pertusa* (Linne) in Trondheimsfjorden on the west coast of Norway, Det Kongelige Norske Videnskabers Selskabs Skrifter 6: 1-9.

Wilson, J.B. 1979. The distribution of the coral *Lophelia pertusa* (L.)(*L. prolifera* (Pallas)) in the north-east Atlantic, Journal of the Marine Biological Association of the United Kingdom 59: 149-164.

Zibrowius, H. 1980. Les Scléractiniaires de la Méditerrananée et de l'Atlantique nord-oriental, Mémoires de l'Institut Oceanographique, Monaco 11: 1-227.

CORAL (*PRIMNOA*) IMPACTED BY FISHING GEAR IN THE GULF OF ALASKA

Kenneth J. Krieger
Auke Bay Laboratory, Alaska Fisheries Science Center
National Marine Fisheries Service, NOAA
Juneau, Alaska, USA

SUMMARY

In 1997 a manned submarine was used in the Gulf of Alaska (GOA) to observe the seafloor where corals (*Primnoa* spp.) had been impacted by fishing gear. A 1990 trawl path was observed where 1 ton of corals had been removed with a research bottom trawl at 365 m depth. Thirty-one coral colonies were observed in the path swept by the trawl net, including 13 large colonies (>1 m^3) and 18 small colonies (<1 m^3). Damage included five large colonies that were missing 95-99% of their branches, and two small colonies that were missing 80% of their polyps. An estimated 27% of the corals in the net path were detached, and 84% of the detached corals were retained in the net. Detached corals that remained on the seafloor were missing 50-90% of their polyps. No young corals had repopulated the colonies damaged seven years previously. Outside the trawl path there were *Primnoa* colonies that had been tipped and dragged. All tipped colonies were attached to small boulders (<0.4 m diameter), and the colonies were tipped in different directions. The colonies may have been snagged by longline gear, which is used extensively in the GOA to catch halibut (*Hippoglossus stenolepis*) and sablefish (*Anoplopoma fimbria*).

INTRODUCTION

The Magnuson-Stevens Fishery Conservation and Management Act places new emphasis on protecting fish habitat. Many commercial fish species are harvested with fishing gear that contacts the seafloor, but we know little about how fishing gear impacts the habitat. One of the main concerns of fishing the seafloor is the damage caused to deep-water corals. Few observations have been made of deep-water corals, and therefore little is known about their biology, ecological role, or how they are impacted by fishing gear. In this study, two trawl paths in the Gulf of Alaska (GOA) where deep-water corals (*Primnoa* spp.) had been removed with a research bottom trawl, were observed from a manned submersible. The main objectives were to determine the proportion of corals removed by the trawl and assess the condition of corals remaining in the trawl path. Also documented were corals outside the trawl paths that had probably been impacted by other types of fishing gear.

Primnoa and other deep-water corals are removed incidentally in the GOA by both fixed and mobile fishing gear that is used to harvest numerous species of fish and shellfish. Bottom trawls are thought to be the most damaging to

Primnoa because this mobile gear contacts large areas of the seafloor. Extensive bottom trawling began in the GOA during the 1960s, by foreign fleets that targeted mainly rockfish. In the 1980s the foreign trawl fleet was replaced by a domestic trawl fleet that uses bottom trawls to target a variety of demersal fish species. How frequently they encounter *Primnoa* is unknown, but *Primnoa* was caught in 168 of 3553 bottom-trawl hauls during the National Marine Fisheries Service (NMFS) GOA and Aleutian surveys in 1990-96 at depths of 100-600 m (catch records from the NMFS bottom-trawl database, Alaska Fisheries Science Center, Seattle, Washington).

Primnoa colonies grow in a treelike form. A calcium carbonate skeleton supports branching colonies of red polyps that capture plankton with stinging cells. Colonies can exceed 2 m in height and 7 m in width (personal observations). In Alaska, *Primnoa* have been reported from Dixon Entrance in southeastern Alaska to Amchitka Island in the Aleutian Islands (Cimberg et al., 1981). *P. willeyi* (Hickson 1915) and *P. resedaeformis* (Gunnerus 1763) have been identified in the GOA, but other *Primnoa* species also probably occur. Distribution records of *Primnoa* indicate these corals extend to 800 m depth and are most abundant in certain inlets in southeastern Alaska and certain bays on the northwestern side of Kodiak and Afognak Islands (Cimberg et al., 1981). *Primnoa* have not been confirmed within SCUBA depths and therefore their minimal depth is probably >30 m. Cimberg et al. (1981) generated a habitat profile for *Primnoa* that includes substrates of large boulders or exposed bedrock, lack of turbidity, and temperatures remaining above 3.7°C. A Norwegian fjord containing *P. resedaeformis* had a temperature range of 4.74-10.62°C and a salinity range of 30.99-34.05‰ (Strömgren, 1970). Cimberg et al. (1981) speculate that *Primnoa* have a limited reproductive period, a life span of more than 100 years, and a growth rate of approximately 1 cm/year. Growth estimates include a 5-cm-diameter piece of *Primnoa* that was isotope-dated at about 500 years (Heikoop et al., 1998) and counts of growth zones that indicate *Primnoa* grow between 1.74 and 2.1 cm per year in height and approximately 0.36 mm in diameter per year (Andrews et al., 2000).

MATERIALS AND METHODS

Study Area

This study was conducted in July 1997 at two sites on the outer continental shelf in the eastern GOA. The sites were 12 km apart and located 25-30 km offshore in Dixon Entrance, which separates southeastern Alaska from British Columbia. *Primnoa* had been removed at the two sites during NMFS bottom-trawl surveys. One site at 365 m depth was trawled for 30 minutes (2.72 km) in 1990 and approximately 1000 kg of *Primnoa* were caught. The second site, at 260 m depth, was trawled for 15 min (1.32 km) in 1996 and approximately 60 kg of *Primnoa* were caught (NMFS bottom-trawl data base, Alaska Fisheries Science Center, Seattle, Washington). Except for the research trawling, the seafloor was unlikely to have been disturbed by other fishing at the 365-m site for several reasons: (i) commercial trawling in the eastern GOA targets schooling rockfish that concentrate at shallower depths, (ii) the seafloor is rugged and presents a high risk of damaging or losing fishing gear, and (iii) large old-growth corals were caught in the 1990 trawl haul. Fishing was more likely to have occurred at the 260-m site because the depth was shallower and the seafloor less rugged.

Trawl Gear

The two sites had been trawled with a polyethylene Nor'eastern trawl rigged with rubber bobbin roller gear (NMFS, 2000). The Nor'eastern net is constructed of 12.7-cm mesh polyethylene web with a 3.2-cm mesh nylon codend liner. The footrope consists of a 1.25-cm chain attached to 0.95-cm-diameter wire. The roller gear is shackled to the footrope and consists of 1.91-cm wire rope with 16, 36-cm-diameter rubber bobbins that are separated by 10-cm-diameter rubber disks. Triple bridles 54.9 m long and made of 1.6-cm-diameter galvanized wire rope connect the wingtips of the net to a pair of steel "V" doors measuring 1.8 m by 2.7 m and weighing 998 kg. The doors travel on the seafloor forward of the net, approximately 15 m outside the wingtips of the net. The bottom set of bridles either contacts the seafloor or travels slightly above the seafloor. The vertical spread of this net ranges from 13 to 18 m between the wingtips, and the horizontal opening ranges from 6 to 10 m between the footrope and headrope.

Submersible

Observations were made from the two-man submersible *Delta*, which is 4.7 m long and battery-powered. It is equipped with halogen lights, video cameras, magnetic compass, directional gyro compass, underwater telephone, and transponder. The transponder allowed charting of the submersible path from a surface vessel. The submersible location was charted at the beginning and end of a dive and at one-minute intervals during a dive using GPS fixes. Halogen lights provided viewing distances of 5-7 m. An externally mounted video camera aimed downward and an internal video camera aimed parallel to the seafloor were used to video-record coral colonies and audio-record observations.

The trawl paths were located from recordings of their latitudes, longitudes and depths. Dives were completed in the net path, in the path swept by the bridles, and outside the path of the net and bridles. During a dive, the pilot maintained the submersible on the seafloor at a speed of 2-3 km/h while the scientist observed through a starboard porthole; external equipment was mounted on the starboard side, and the side portholes provided the widest range of view. The pilot sat above the observer in a tower with a panoramic view and documented corals on the port side. Colonies were approached for detailed observations and for video-recording, and a loop was completed around the perimeter of large colonies. The volume of each colony was determined from length, width and height estimates. Size references for most estimates were 20-cm laser beams projected onto the seafloor and recorded onto the video

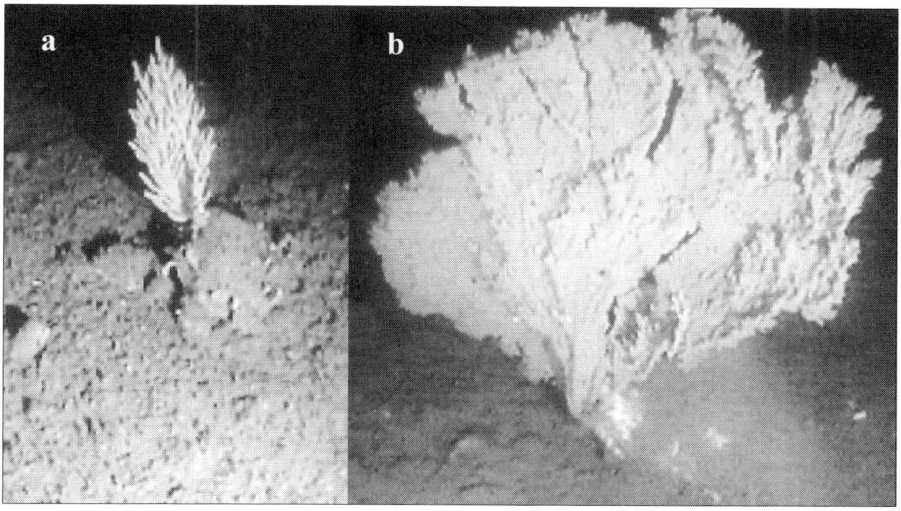

*Figure 1. A small colony (a) and large colony (b) of coral (*Primnoa *spp.) observed from a submersible in the Gulf of Alaska, July 1997*

tape. The length and height of the submersible were used to estimate the dimensions of extremely large colonies. Sizes were estimated to within 0.3 m for large coral colonies (volumes 1-20 m^3) and to within 0.1 m for small coral colonies (volumes <1 m^3). Diameters of boulders with corals attached were estimated to within 0.3 m for boulders >1.0 m and to within 0.1 m for boulders <1.0 m. Coral trunks and white calcium deposits remained on boulders where corals had been removed. The volume of these removed corals was estimated from the average volume of intact colonies with similar-diameter trunks. Branches of coral lying on the seafloor were counted, and estimates made of their volume and the percent of polyps missing from each branch.

RESULTS

Substrate at the 365-m site was pebble and cobble 2-25 cm in diameter with scattered boulders. Boulders were usually in the field of view, and large boulders (0.5-3.0 m diameter) were common. Substrate at the 260-m site was sand and pebble with scattered cobble and boulders. Boulders >0.5 m diameter were rare. *Primnoa* colonies at the two sites ranged in volume from 0.01 m^3 to 17.14 m^3 and were attached to <5% of the boulders. Colonies consisted of single trees or multiple trees. Small colonies were 0.1-1.5 m high, with sparse branches (Figure 1), and large colonies were 1-3 m high, with numerous secondary branches that occupied most of the volume between major branches (Figure 1).

1990 Trawl Site

The 1990 net path at 365-m depth was identified by displaced boulders, broken corals, and pieces of trawl web. The net path was observed for 0.68 km of the 2.72 km total distance and 31 coral colonies were counted, including 13 large colonies and 18 small colonies. Damage included five large colonies that were missing 95-99% of their branches, and two small colonies that were missing 80% of their polyps (Table 1). The estimated total volume of the large colonies was 59.4 m^3, of which 16.1 m^3 (27%) was missing. The estimated total volume of the small colonies was 3.0 m^3, of which 0.5 m^3 (17%) was missing polyps. The damaged large colonies were attached to boulders 1.3-2 m in diameter that had been tipped 45°-90°, and four of these boulders had been dragged 1-3 m (Figure 2). The damaged small colonies were attached to boulders 0.3 m and 1.3 m in diameter, and both boulders were tipped about 90°. None of the boulders without coral appeared to have been dragged or tipped. Eight unattached branches were lying in the net path (Figure 3), and six of the branches were within 3 m of the damaged colonies. The estimated volume of the eight branches was 3.5 m^3, 18% of the coral removed from boulders. Between 50% and 99% of the polyps were missing from these branches (Table 1).

The area swept by the bridles was observed by transecting perpendicular to the trawl path. Approximately 50 coral colonies were observed within

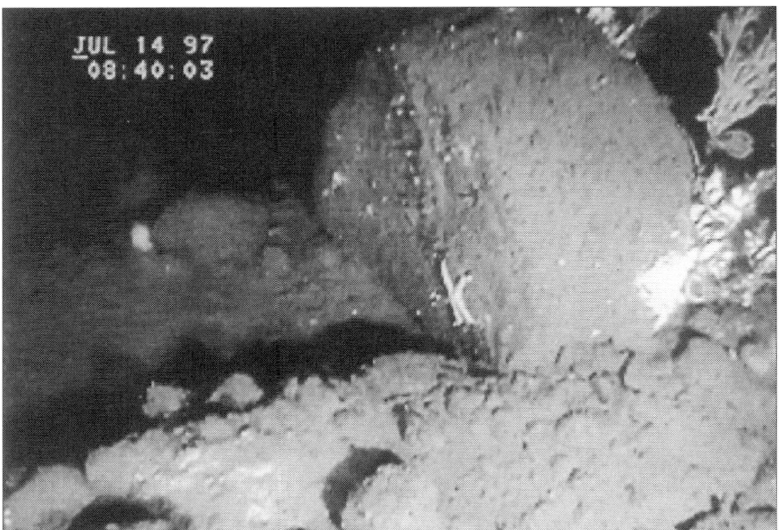

*Figure 2. A 2-metre-diameter boulder that was dragged and had most of the coral (*Primnoa *sp.) removed from it by a bottom trawl. Observation was from a submersible in the Gulf of Alaska, July 1997.*

10 m of the net path where the bridles had swept the seafloor; no colony was damaged, and no boulder had been disturbed.

Five dives were completed between 1 km and 5 km of the 1990 trawl path. These dives traversed 4.88 km of seafloor, and 234 coral colonies were observed, including 173 small and 61 large colonies. None of these colonies or boulders had been disturbed and no colony had been removed from boulders, indicating that the 1990 trawling was the only fishing that had disturbed the habitat in this area.

1996 Trawl Site

The 1996 net path at 260-m depth was identified by depressions on the seafloor caused by the rubber bobbins attached to the footrope. The net path was followed for 0.30 km of the 1.32 km total distance; seven small colonies, two large colonies, and five detached branches were observed. All nine colonies appeared to have been damaged by fishing gear, but some damage had occurred before the 1996 trawling because (i) colonies were tipped in different directions, (ii) bryozoans and other invertebrates were more established on newly exposed surfaces of tipped boulders in the 1996 trawl path compared to newly exposed surfaces of tipped boulders in the 1990 trawl path, and (iii) corals had repopulated some of the calcium bases where corals were missing. Therefore, the specific damage caused by the 1996 trawling could not be determined because of previous fishing activity.

*Figure 3. Coral (*Primnoa *sp.) that was detached from a boulder by a bottom trawl 7 years previously. Observation was from a submersible in the Gulf of Alaska, July 1997.*

A single transect was completed within 5 km of the 1996 trawl path, and 69 small colonies and nine large colonies were observed. Nineteen small colonies were tipped, and three of them had been dragged <1 m (Figure 4). The colonies were tipped in different directions, and all tipped colonies were attached to boulders <0.4 m in diameter. Two anemones (species unknown) that were attached to <0.3-m boulders were also tipped and dragged. These corals and anemones were probably not tipped by bottom trawls, because only small boulders with invertebrates were tipped, no large colonies had been tipped or damaged, and no broken corals were observed. These invertebrates may have been snagged by longline gear used extensively in the GOA to catch halibut (*Hippoglossus stenolepis*) and sablefish (*Anoplopoma fimbria*). Longline gear consists of a 0.5-1.0-cm mainline anchored at both ends with hooks and leaders attached every 1-4 m. High (1998) observed the retrieval of longline gear from a submersible and observed coral branches being snagged by hooks during gear retrieval. *Primnoa* and other coral species were caught on 619 of 541,350 hooks that were fished during the sablefish longline survey in the

Table 1. Damage to coral (*Primnoa* spp.) and movement of their boulder bases by a bottom trawl. Observations were from a submersible in the Gulf of Alaska in 1997.

Large Coral Colonies				
			Boulder base	
Colony number	Size (m^3)	Removed (%)	Diameter (m)	Movement
1	4.2		2.0	
2	3.1		1.6	
3	2.6		1.0	
4	3.6		1.3	
5	3.1		1.6	
6	3.6		1.6	
7	1.7		1.6	
8	4.1		2.0	
9	17.1		3.0	
10	3.3	95	2.0	3.0 m
11	3.3	99	2.0	2.0 m
12	3.3	99	1.6	tipped
13	3.3	99	2.3	3.0 m
14	3.3	95	1.3	1.5 m

Small Coral Colonies				
			Boulder base	
Colony number	Size (m^3)	Dead (%)	Diameter (m)	Movement
1	0.1	80	0.3	tipped
2	0.0		0.6	
3	0.1		1.0	
4	0.1		1.3	
5	0.1		1.3	
6	0.0		1.6	
7	0.5	80	1.3	tipped
8	0.1		1.6	
9	0.0		0.3	
10	0.1		1.0	
11	0.4		1.6	
12	0.2		1.3	
13	0.0		0.3	
14	0.2		1.3	
15	0.6		0.6	
16	0.1		0.6	
17	0.0		1.3	
18	0.2		1.6	

Upper left. *The CAMPOD video/photo platform being deployed over the side of the C.C.G.S.* Hudson. ***Centre left.*** *The soft coral, Alcyonacea-type 1, on coraline-encrusted cobbles at the Stone Fence.* ***Lower left.*** *The gorgonacean* Primnoa resedaeformis, *one of the most common and abundant corals off Atlantic Canada.* ***Upper right.*** *An intact specimen of* P. resedaeformis *recovered by the videograb at Romeys Peak in the Northeast Channel.* ***Centre right.*** *Blue form of the species group Alcyonacea-type 2 growing on a boulder in the Gully.* ***Lower right.*** *Solitary and colonial anemones observed commensal on the branches of the gorgonacean coral Gorgonacea-type 1.*

Crushed and scattered Lophelia pertusa *on part of the Norwegian continental shelf where reefs have been damaged by bottom trawling.*

Photograph courtesy of Dr. Jan Helge Fossa.

Lophelia pertusa *reef on the Norwegian continental shelf at 260m depth. The living coral is white; dead skeletal remains are darker. The red fish are* Sebastes *sp.*

Photograph courtesy of Dr. Jan Helge Fossa.

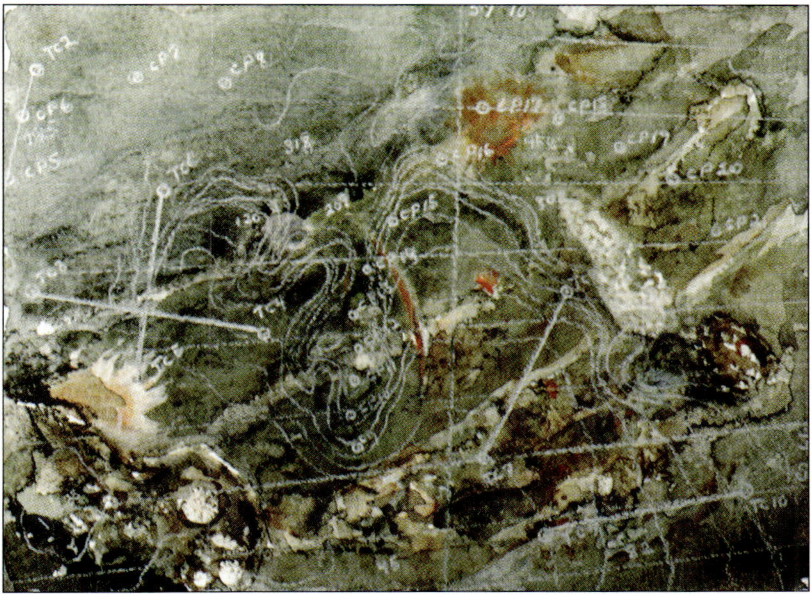

Haldemand Canyon, Yr. 1999, watercolour, colour pencil, on paper, 20 x 26 cm.

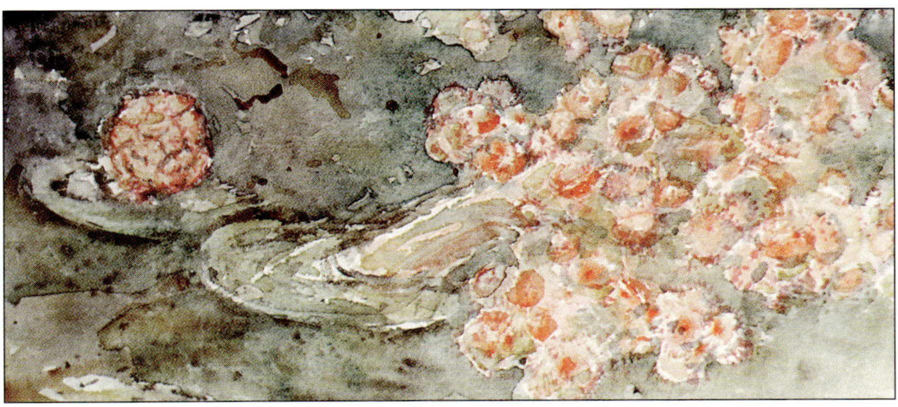

Gersemia rubiformis. *Yr. 1999, watercolour on paper, 17 x 38 cm*
Artist Susan Feindel. For more information, see paper by Susan Feindel.

Primnoa, *Yr. 1999. Watercolour on paper, 15 x 20 cm.*

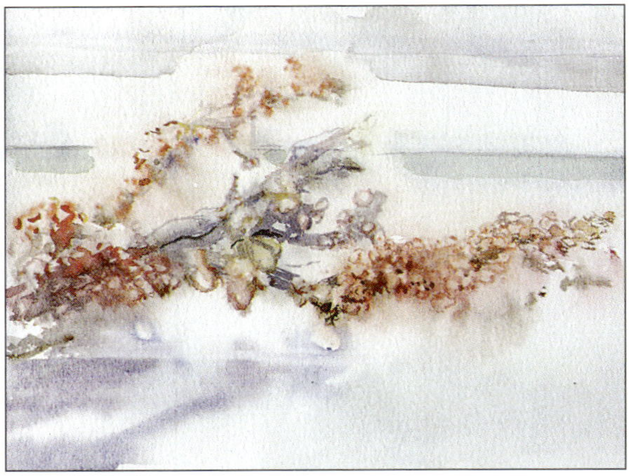

When Human Activity Subsides, Life Blooms From Industrial Waste. Yr. 2000, oil on canvas, brass trailings, acrylic paint, 81x91 cm. Artist Susan Feindel. For more information, see paper by Susan Feindel.

Paragorgia arborea *and* Primnoa resedaeformis *growing on a boulder in the Northeast Channel region of the Scotian Shelf (eastern Canada) at about 450 m. Numerous redfish (*Sebastes *spp.) aggregate around these large gorgonian corals.*

ROPOS submersible sampling Paragorgia arborea *in Northeast Channel.*

Pandalid shrimps among polyps of Primnoa resedaeformis *in the Northeast Channel, eastern Canada.*

A boulder bearing Primnoa resedaeformis *that has been rolled on its side. A basketstar (*Gorgonocephalus *sp.) is at the right, and a rope is in the upper left corner. Northeast Channel at about 450m.*

Close-up of polyps of Paragorgia arborea*, with tentacles extended. Photographed* in situ *in Northeast Channel.*

A small anemone commonly colonizes the skeletons of the coral Primnoa resedaeformis, *as shown here.* Primnoa *polyps are absent in this example, but may be present on parts of the same skeleton.*

A boulder reef at about 35 m near Cape Sable, Nova Scotia. The boulder is covered with the small "soft coral" Gersemia rubiformis *(Alcyonacea).*
Photo by John Batt, Dalhousie University.

Gorgonian corals in the Northeast Channel, Nova Scotia. Note that a large dead Paragorgia arborea *("bubblegum tree" as named by fishermen) is lying among the living corals, perhaps having been pulled over by fishing activity.*
Copyright ROPOS 2001, reproduced with permission.

*Figure 4. Coral (*Primnoa *sp.) on a 40-cm-diameter boulder that had been dragged and tipped, probably by longline fishing gear. Observation was from a submersible in the Gulf of Alaska, July 1997.*

*Figure 5. Corals (*Primnoa *sp.) 5-40 cm high that were attached to cable. Observation was from a submersible in the Gulf of Alaska, July 1997.*

GOA and Aleutian Islands in 1998 at 150-900 m depths (NMFS sablefish longline database, Auke Bay Laboratory, Juneau, Alaska).

Coral Recovery

Young colonies <20 cm high were not observed in the 1990 trawl path, indicating that corals had not recolonized the sites where corals had been removed seven years previously. Six young colonies were observed in the 1996 trawl path, four of which were growing on calcium bases where corals had been removed before 1996. Young colonies were also attached to two cables adjacent to the 1990 trawl path (Figure 5). These cables were approximately 7 cm in diameter and 5 m long, and each contained >15 individual colonies that ranged from 5 to 40 cm high. The type of cable was not determined because it was covered with invertebrates and silt.

DISCUSSION

A single trawl haul is assumed to have caused the damage to corals at the 365-m site because the seafloor had not been disturbed adjacent to the 1990 trawl path. The estimates of coral volume damaged by a single trawl haul are not precise because they are based on the "average" size of colonies that were not removed. The number of colonies damaged and how they were damaged are more precise estimates of the trawling impact. The 1990 trawl haul damaged mainly large colonies by removing most of the branches from 5 of 13 colonies. The only boulders that had been moved were those with damaged colonies, indicating that the colonies became entangled in the net and the forward movement of the net tipped the boulders until the colonies detached from the boulders. The Nor'eastern trawl appeared to cause minimal damage to small colonies, perhaps because they contain smaller and fewer branches that are less likely to entangle in the net. *Primnoa* can apparently pass under trawl gear without breaking because they are strong and flexible; the submersible pushed against *Primnoa* on several occasions and the branches bent without breaking. Damaged corals were not observed where the bridles swept next to the net, but detailed observations are needed of the entire sweep of the bridles to determine their impact on corals.

In the 1990 net path, broken corals that remained on the seafloor and small colonies that were tipped were missing at least 50% of their polyps. Polyps had probably died because they could not feed effectively, compared to corals that are upright and perpendicular to currents. Most young corals were growing on old coral bases or on cable, indicating that the settling of planulae larvae is substrate-limited. The wide range of coral sizes on the cable indicated larvae had settled during multiple years. The 1990 trawling had created calcium bases, but planula larvae had apparently not settled on them.

Thousands of kilometres of longline gear with hooks are deployed each year in the GOA, and some hooks snag corals and other invertebrates. High (1998) noted that unless lifted off bottom, sturdy flexible corals appeared to be relatively unharmed by longline components. My observations indicate that small coral colonies and other invertebrates attached to small boulders may be pulled over when snagged by longline gear. In British Columbia, large *Primnoa* colonies on a steep ridge were tipped by strong currents (Tunnicliffe and Syvitski, 1983). The latter study noted that tipped corals remaining alive had some branches growing at 90° bends and orienting vertically into the current.

Shallow-water corals have been studied extensively using SCUBA, and they are known to provide habitat and prey for many types of fish and invertebrates and share a wide diversity of symbiotic relationships with other species (see Jones and Endean, 1973-1976, for an overview). The limited information on deep-water corals indicate they have a similar ecological importance. Jensen and Frederiksen (1992) identified 298 species associated with the deep-water coral *Lophelia pertusa*, and Mortensen et al. (1995) found a high density and diversity of fauna associated with *L. pertusa*. Some GOA areas contain populations of *Primnoa* and other deep-water corals that have not been impacted by fishing gear because they are among rugged habitats, such as those found on the shelf break (personal observations). These habitats, however, are becoming more vulnerable to fishing because of improved fishing technology, including larger vessels, improved electronic equipment, and gear that can fish deeper and more rugged habitats. Areas containing deep-water corals need to be identified and protected because they are a unique habitat and if damaged, will take decades and possibly centuries to recover.

REFERENCES

Andrews, A.H., E.Cordes, J. Heifetz, G.M. Cailliet, K.H. Coale, D. Somerton, K. Munk and M. Mahoney. Age and growth of a deep-sea, habitat-forming octocorallian (*Primnoa* sp.) from the Gulf of Alaska, with radiometric age validation, Hydrobiologia (in press).

Cimberg, R.L., T. Gerrodette and K. Muzik. 1981. Habitat requirements and expected distribution of Alaska coral. Final Report, Research Unit 601, VTN Oregon, Inc. U.S. Department of Commerce, NOAA, OCSEAP Final Report 54 (1987), pp. 207-308. Office of Marine Pollution Assessment, 701 C Street, Anchorage, Alaska 95513.

High, W.L. 1998. Observations of a scientist/diver on fishing technology and fisheries biology. U.S. Department of Commerce, National Marine Fisheries Service, Alaska Fisheries Science Center Processed Report 98-01, Seattle, Washington.

Heikoop, J.M, M.J. Risk, A.V. Lazier and H.P. Schwarcz. 1998. $\partial^{18}O$ and $\partial^{13}C$ signatures of a deep-sea gorgonian coral from the Atlantic coast of Canada. Spring American Geophysical Union, Boston Meeting. Abstract.

Jensen, A. and R. Frederiksen. 1992. The fauna associated with the bank-forming deep-water coral *Lophelia pertusa* (Scleractinaria) on the Faroe shelf, Sarsia, 77: 53-69. Cited in Mortensen et al. (1995).

Jones, O.A., and R. Endean (eds). 1973, 1974, 1976. Biology and Geology of Coral Reefs. Vols. I, II, and III. Academic Press, New York.

Mortensen, P.B., M. Hovland, T. Brattegard and R.A. Frestveit. 1995. Deep water bioherms of the scleractinian coral *Lophelia pertusa* (L.) at 64° N on the Norwegian shelf: structure and associated megafauna, Sarsia 80: 145-158.

NMFS (National Marine Fisheries Service). 2000. ADP Code Book. U.S. Department of Commerce, Resource Assessment and Conservation Engineering Division, NMFS, Alaska Fisheries Science Center, Seattle, WA, 100 pp.

Strömgren, T. 1970. Emergence of *Paramuricea placomus* (L.) and *Primnoa resedaeformis* (Gunn.) in the inner part of Trondheimsfjorden (west coast of Norway), Det Kongelige Norske Videnskabers Selskab Skrifter 4: 1-6.

Tunnicliffe, V. and J.P.M. Syvitski. 1983. Corals move boulders: an unusual mechanism of sediment transport, Limnology and Oceanography 28: 564-568.

PROTECTING GORGONIAN CORALS OFF ALASKA FROM FISHING IMPACTS

David Witherell and Cathy Coon
Staff, North Pacific Fishery Management Council
Anchorage, Alaska, USA

SUMMARY

Large deep-sea corals of the Order Gorgonacea are found in the North Pacific Ocean off Alaska, USA. The North Pacific Fishery Management Council has identified these corals as essential fish habitat of particular concern, so management measures to reduce the fishery impacts are being considered. These corals have been shown to be (i) important shelter for rockfish and other fish species, (ii) very long-lived, (iii) easily damaged by fishing gear, and (iv) slow to recover from damage. Coral conservation measures previously implemented include trawl closure areas and a prohibition on commercial harvests. Because other fishing gears (e.g., longlines and pots) also impact corals, a prohibition of all gear types in areas of high coral concentration was proposed and evaluated. The proposal was tabled because available scientific data on coral distribution was at too large a scale to define discrete locations of coral colonies. Additionally, many fishermen using longline and pot gear were concerned about being displaced from areas they had previously fished, and many felt that their gear caused less damage to corals than does trawl gear. Involvement of stakeholders at the local level will allow for better information exchange, including information on coral distribution, fishing gear impacts, and development of appropriate management measures.

INTRODUCTION

It is not widely known that corals commonly occur in the cold waters off the Alaska coast of the United States. The focus of national and international legislation designed to protect corals has been on shallow-water coral reefs located in tropical areas. Protection of cold-water corals has been generally overlooked, yet these corals are no less spectacular from a biological, ecological and even aesthetic standpoint. Perhaps it is just a case of 'out of sight, out of mind.' The purpose of this paper is to review the need to protect coral from fishing impacts, and to discuss an approach to achieving this objective.

The largest and perhaps most charismatic of the deep-sea corals belong to the Order Gorgonacea. Gorgonians are colonies of animals composed of individual polyps which deposit a tree-like skeleton. Common gorgonians off Alaska include red tree coral (*Primnoa willeyi* and *P. resedaeformis*), bubblegum coral (*Paragorgia arborea*) and other sea fans (*Calligorgia* sp.) (Heifetz, 1999). These large coral colonies (up to 3 m high and 7 m wide) are slow growing and may be very long-lived (>500 years) (Risk et al., 1998). The colonies are attached to rocks and generally occur in discrete aggregations (like groves of trees) in deep water (10-800 m) characterized by fairly strong currents (3-4 knots) and low turbidity (Cimberg et al., 1981; Krieger, 1998). When alive, deep-sea gorgonian corals are brightly coloured and make for breathtaking underwater sights. There are many reasons to protect large deep-sea corals, some of which are related to their ecological functioning, others to their use by mankind.

Corals create complex habitat by providing shelter, prey and habitat for fish. Fosså et al. (1999) observed dense aggregations of rockfish (*Sebastes* sp.) on deep-water scleractinian coral (*Lophelia* sp.) off Norway. Longline catches of rockfish and other species were much higher in coral areas than in surrounding areas without corals. Other studies in Norway have also shown the importance of deep-water corals in marine ecosystems, in providing habitat and prey for other megafauna species (Mortensen et al., 1995). Krieger (1999) reported that rockfish were associated with *Primnoa* and that starfish were preying on the colonies.

Deep-sea corals also have use for medicine and scientific research. Sea fans are known to contain high concentrations of prostaglandins, a 'wonder drug' used to treat heart disease and asthma (Faulkner, 1992). Corals also contain pseudopterosins (a pain killer) and gorgonians produce antibiotics. Another scientific use of corals is as a time record of sea temperatures. Skeletons of *Primnoa* colonies grow directly in relation to water temperature so they can provide a historical record of ocean temperature (Smith et al., 1997). A 5 cm diameter specimen of *Primnoa* was estimated to be about 500 years old, based on isotope dating (Risk et al., 1998). Given that larger specimens have been reported, *Primnoa* colonies may provide temperature data from the last thousand years or more, thus allowing scientists to test hypotheses on climate change and regime shifts.

In the United States, protection of gorgonian corals may be warranted under the Magnuson-Stevens Act (the law that establishes the U.S. national program for conservation and management of fishery resources), as modified by the Sustainable Fisheries Act of 1996. The law requires fishery management councils to minimize, to the extent practicable, adverse effects on essential fish habitat caused by fishing, and identify other actions to encourage the conservation and enhancement of such habitat (Fluharty, 2000). Areas of special importance that may require additional protection from adverse effects were designated as habitat areas of particular concern. Habitat areas of particular concern are defined on the basis of their ecological importance, sensitivity, exposure, and rarity of the habitat. Gorgonian corals epitomize of this type of special habitat.

Where Are Corals Located Off Alaska?

Gorgonian corals are widely distributed along the continental shelf and slope along the Gulf of Alaska and the Aleutian Islands. Distribution information was examined using two databases: the National Marine Fisheries Service bottom trawl research surveys, and incidental catch data collected by fishery

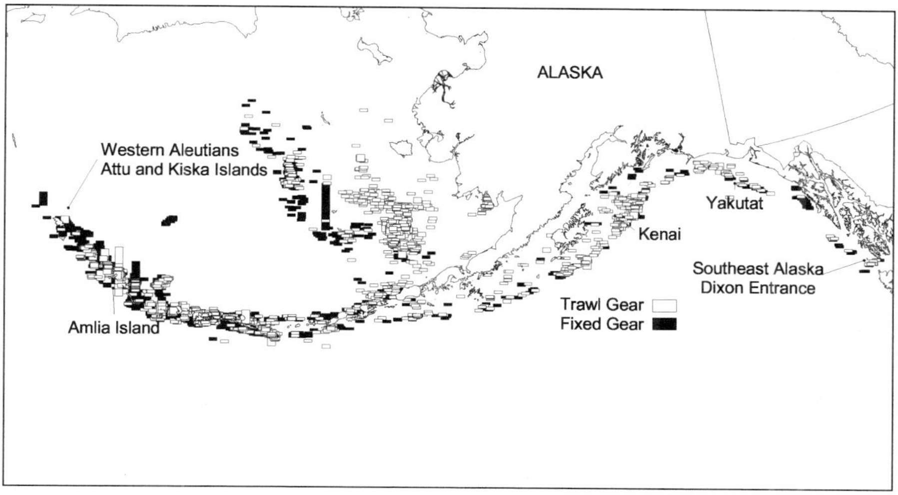

Figure 1. Relative abundance and distribution of gorgonian corals based on trawl survey data, 1954-1998

observers. Gorgonian corals off coastal Alaska were analyzed from the 1954-1998 survey data and plotted by location. Catch per unit effort was calculated by weight of gorgonians (kg) in trawl by the area swept (distance towed * net width) for each tow with coral present. Observer data included any haul containing coral from any species during the period 1991-1998. Unfortunately, there is no taxonomic identification of coral in the observer database, and coral is combined with bryozoans in the records, so this database has only limited use as supplemental information on distribution.

The relative distribution of gorgonian coral from survey data is shown in Figure 1. Relative total amounts of gorgonian coral were highest off southeast Alaska near the Canadian border, at the western end of the Aleutian Islands, and near the Kenai Peninsula. The highest CPUEs for gorgonians occurred near Attu Island, Kiska Island, Amlia Island, the Kenai Peninsula, Yakutat and Dixon Entrance.

Relative distribution of coral and bryozoans using observer data of trawl, longline and pot gear fisheries is shown in Figure 2. The relative longline and pot gear catch of coral (all species) in the Aleutian Islands was highest in the vicinity of Kiska, Attu and Amchitka Islands. The highest observed catch in trawl fisheries was taken in the Aleutian Islands in the vicinity of Kiska Island and the Pribilof Islands, but analysis of survey data indicated that the Pribilof catch was likely all bryozoans (NPFMC, 2000). No catch of coral of a similar magnitude was found in trawl catches in the central Gulf of Alaska nor in southeast Alaska. The only relative catch in fixed gear fisheries of a similar magnitude to that found in the Aleutian Islands was found off southeast Alaska.

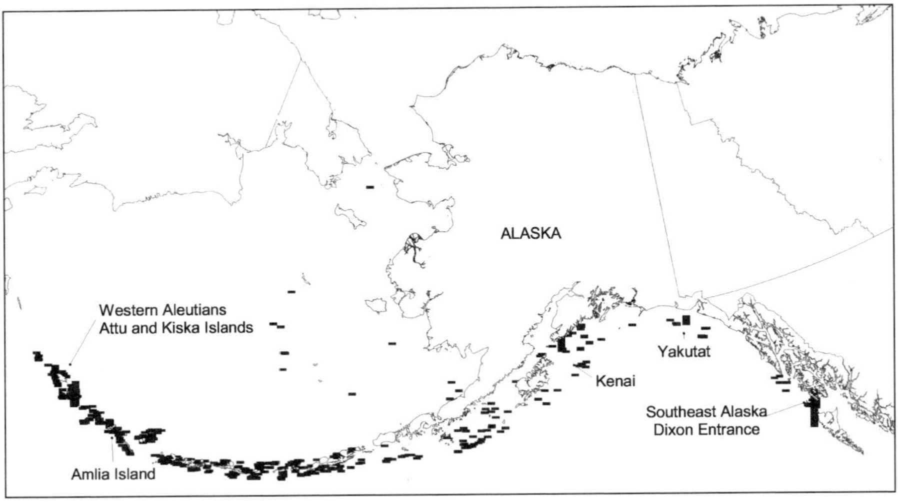

Figure 2. Relative abundance and distribution of all corals and bryozoans based on commercial fishery observer data, 1991-1998

How Does Fishing Gear Affect Corals?

Very few studies have been conducted on the effects of fishing gear on deep-sea corals. Research has shown that trawl gears can damage sponges and hard corals on tropical shallow-water reefs (Tilmant, 1979; van Dolah et al., 1987; Sainsbury et al., 1997) and sponges off Alaska (Freese et al., 1999). Much less is known about the effects of fishing on deep, cold-water corals. Until recently, all that was known for certain was that in areas of fishing, corals had become less abundant (Breeze et al. 1997). In the past couple of years however, biologists have become concerned about fishing impacts on these deep-water corals and studies are underway around the world where these corals exist.

Three recent studies have shows that deep-water corals are impacted by trawling, even at low intensity. Underwater observations of deep-water (200-400 m) scleractinian coral (*Lophelia* sp.) reefs off Norway have confirmed that trawl gear caused considerable damage (Fosså et al., 1999). A remote-operated vehicle, combined with an accurate underwater tracking system, was used to quantify damage caused by shrimp trawls off Australia (Pitcher et al., 1999). They found that about 10% of the fan gorgonians were removed with each pass of the trawl. Krieger (1998) used a submersible to examine the damage of *Primnoa* caused by a single pass of a research trawl off Dixon Entrance, Alaska, at 365 m depth, made seven years earlier. The net had removed approximately 1000 kg of *Primnoa*. He observed that about 30% of the coral in the trawl path was removed or broken.

Few observations have been made on the effects of other fishing gear types such as longlines and pots. At a 260 m site off Alaska, rocks and small boulders with attached *Primnoa* and sea anenomes that had been tipped and dragged, probably by longline gear used to fish halibut (*Hippoglossus stenolepis*) and sablefish (*Anoplopoma fimbria*) (Ken Krieger, NMFS, pers. comm.). High (1998) observed longline gear from a submersible and noted that large branches of corals were snagged by loose groundlines or hooks, and that the hard corals (i.e., *Primnoa* sp.) often had portions broken off. Longlines also catch and retain gorgonians and other corals off Alaska, based on observations from the NMFS sablefish longline survey (Jon Heifetz, NMFS, pers. comm.), the commercial fisheries observer program, and anecdotal reports from fishermen. Fishermen off Nova Scotia report that longline gear gets tangled up and catches coral when gear is set in areas of coral "trees" (Breeze et al., 1997). The Alaska golden king crab pot fisheries sometimes snag gorgonian corals in the Aleutian Islands area, and observers are now beginning to identify and record these incidental catches (L. Byrne, Alaska Department of Fish and Game, pers. comm.). An anecdotal report cited in Risk et al. (1998) describes the bycatch of red tree corals entangled in mesh of prawn pots off British Columbia, Canada, and the subsequent disappearance of corals in only six years.

What Has Been Done to Date to Protect Coral off Alaska?

Marine protected areas and catch controls are the two types of management measures have helped to protect deep-sea corals from fishing impacts off Alaska. The government bodies established to manage the groundfish fisheries (North Pacific Fishery Management Council and National Marine Fisheries Service) have implemented marine protected areas to protect sensitive bottom habitats and to reduce potential competition of the fisheries with marine mammals (Witherell et al., 2000). In most cases, only mobile gear (trawls, dredges) has been excluded from the marine protected areas.

In 1991, longline fishermen from Sitka and other local citizens proposed that all trawling be prohibited off southeast Alaska. The rationale for this was that trawling was causing long-term damage to deep-sea corals, causing conservation problems for rockfish, and causing social disruption to the local fishing industry (Behnken, 1993). Although the original proposal was not adopted when brought forth to the Council for final decision, it was later adopted as part of another amendment to the groundfish fishery management plan. Beginning in 1998, all trawling was prohibited in southeast Alaska, a 52,600 nmi^2 area.

A more recent initiative by concerned citizens of Sitka led to establishment of the Sitka Pinnacles Marine Reserve. All bottom-fishing gear types (except pelagic troll gear used for salmon) were prohibited in this small 3.1 nmi^2 area. These pinnacles contain high-relief habitat, including red tree corals, where lingcod and several species of rockfish aggregated. Increased effort by the charter boat and commercial fleet raised concerns about localized depletion of lingcod and associated habitat impacts (O'Connell et al., 1998). There was strong local support for this reserve as a result of extensive research, underwater video, and public outreach.

In April 2000, the Council adopted a regulation that defines all corals and sponges as prohibited species. The purpose of this amendment was to prohibit a full-scale commercial fishery from developing on invertebrates that provide important habitat for fish (NPFMC, 2000). Because the skeleton can be cut and polished to a lustre, gorgonian corals had previously been harvested commercially off Alaska and sold for jewelry (Cimberg et al., 1981). The regulation will prohibit the sale, barter and trade of corals and sponges, but allow retention for personal use. Gorgonian corals are often retained as souvenirs by fishermen when caught incidentally in fisheries off Nova Scotia (Breeze et al., 1997) and off Alaska (Dan Falvey, commercial longline fishermen, pers. comm.).

What Else Can We Do to Protect Deep-sea Corals?

Because gorgonian corals don't movè, establishment of a marine reserve or marine protected area in coral "hotspots" seems to be one reasonable management option worth consideration. Marine protected areas have been widely supported in the scientific literature (e.g., Agardy, 1994; Allison et al., 1998; Lauck et al., 1998; Lindeboom, 2000) as a management tool to protect biodiversity, benthic habitats, viable populations, and ecological processes. Marine protected areas implement the precautionary approach by reducing risk and hedging against uncertainties, errors and biases in fisheries management, thus providing insurance against fishery collapses. In Alaska, marine protected areas have been implemented to prevent damage to vulnerable living substrates caused by trawl gear (Ackley and Witherell, 1999; Witherell et al., 2000). United States President Clinton recently issued an Executive Order to establish a national system of marine protected areas — known as New Ocean Conservation Zones — where all fishing, oil drilling, and other consumptive uses of marine resources are prohibited, to preserve marine resources and ensure sustainable economic use of the ocean

The objective of establishing marine protected areas where gorgonian coral colonies aggregate would be to protect this vulnerable benthic habitat from potential degradation due to fishing, and reap all other benefits provided by such protection. In this particular case, any fishing or non-fishing activity that adversely impacts gorgonian corals should be tightly regulated within these marine protected areas. The most conservative approach would prohibit all bottom-fishing gear from the areas, as it is likely that all fishing gear that touches corals would cause some damage. Fisheries using longline, pot, dredge and trawl gear, however, would be impacted by such action and the social and economic impacts could be significant depending on the size and location of the designated closure areas.

In February 2000, the Council reviewed a bio-economic impact analysis of a proposal to protect gorgonian corals by establishing six marine protected areas, totaling over 7,000 nmi^2. After considerable review and debate, the proposal was tabled for several reasons. First, some fishermen felt that the proposed areas were too large and did not reflect the patchiness of gorgonian coral colonies. Second, quantitative scientific data are limited on coral distribution, so closure areas based on just survey data may divert fishing effort into other areas not currently known to contain corals. Third, many fishermen using longline and pot gear were concerned about being displaced from areas they had previously fished, and many believed that their gear caused less damage to corals than trawl gear. The Council decided that many of these concerns could be addressed by increasing the involvement of coastal community stakeholders in the development of appropriate management measures to protect gorgonian coral colonies.

In this case, the involvement of local community stakeholders, rather than just relying on regional fishing industry spokespersons, is critical. Coastal community stakeholders can provide local and traditional knowledge regarding the abundance and location of benthic habitat types. This information can be extremely useful and valuable in cases where data are limited (e.g., gorgonian corals). Additionally, with increased understanding, stakeholders may take more responsibility for habitat protection and regulatory compliance, thereby reducing the need for enforcement (Lindeboom, 2000). Enforcement has been a major impediment in the effectiveness of marine reserves, particularly when they are small in size or apply to only certain gear types. In the North Pacific region, coastal community stakeholders include not only individuals such as commercial fishers, recreational fishers and other local people, but also larger groups and coastal businesses involved in fish processing, mining and dredging, and others.

The stakeholder approach we will be using is based on the focus-group meeting format recommended by Pederson and Hall-Arber (1999) for collecting fish habitat information from fishermen. Our plan is to hold informal meetings in coastal communities to disseminate information, gather local knowledge of coral distribution, build consensus on the need to protect these corals, and develop appropriate management alternatives. We expect this to be a difficult task, but hope that our efforts will provide valuable feedback and insights to protect gorgonian corals from fishing impacts.

REFERENCES

Ackley, D. and D. Witherell. 1999. Development of a marine protected area in Bristol Bay, Alaska. Ecosystem Approaches for Fisheries Management. Alaska Sea Grant College Program AK-SG-99-01.

Agardy, M.T. 1994. Advances in marine conservation: The role of marine protected areas, Trends in Ecology and Evolution 9: 267-270.

Allison, G., J. Lubchenco and M. Carr. 1998. Marine reserves are necessary but not sufficient for marine conservation, Ecological Applications 8(1): S79-S92.

Behnken, L. 1993. Southeast Alaska trawl closure: A case study in risk-averse management, Sea Wind 7(1): 8-14.

Breeze H., D.S. Davis, M. Butler and K. Vladimir. 1997. Distribution and status of deep sea corals off Nova Scotia. Marine Issues Committee Special Publication Number 1, Ecology Action Centre, Halifax, N.S., 58 pp.

Cimberg R.L., T. Gerrodette and K. Muzik 1981. Habitat requirements and expected distribution of Alaska coral. Final Report to Office of Marine Pollution Assessment, Alaska Office. Research Unit 601.

Faulkner, D.J. 1992. Biomedical uses for natural marine chemicals, Oceanus 35: 29-35.

Fluharty. D. 2000. Habitat protection, ecological issues, and implementation of the Sustainable Fisheries Act, Ecological Applications 10(2): 325-337.

Fosså, J.H., D.M. Furevik, P.B. Mortensen and M. Hovland. 1999. Effects of bottom trawling on *Lophelia* deep water coral reefs in Norway. Institute of Marine Research, Bergin, Norway. Poster presented at ICES meeting on Ecosystem Effects of Fishing, March 1999, Montpellier, France.

Freese, L., P.J. Auster, J. Heifetz, and B.L. Wing. 1999. Effects of trawling on seafloor habitat and associated invertebrate taxa in the Gulf of Alaska, Marine Ecology Progress Series 182: 119-126.

Heifetz, J. 1999. Description and distribution of coral in the Gulf of Alaska and Bering Sea: a progress report, in P. Livingston (ed.), Ecosystem Considerations for 2000. North Pacific Fishery Management Council, Anchorage, Alaska, pp. 44-45.

High, W.L. 1998. Observations of a scientist/diver on fishing technology and fisheries biology. Alaska Fisheries Science Center, Seattle, processed report 98-01.

Krieger, K. 1998. *Primnoa* spp. observed inside and outside a bottom trawl path from a submersible. Abstract. 10th Western Groundfish Conference, p. 88.

Krieger, K.J. and B. Wing. Megafauna associations with gorgonian corals (*Primnoa* sp.) in the Gulf of Alaska. (Unpublished manuscript)

Lauck, Y., C.W. Clarck, M. Mangel and G.R. Munro. 1998. Implementing the precautionary principle in fisheries management through marine reserves, Ecological Applications 8(1): S72-S78.

Lindeboom. H.J. 2000. The need for closed areas as conservation tools. In M.J. Kaiser and S.J. deGroot (eds), The Effects of Fishing on Non-target Species and Habitats, Blackwell Science Ltd, Oxford, pp. 290-301.

Mortensen P.B., M. Hovland, T. Brattegard and R. Frestveit. 1995. Deep water bioherms of the scleractinian coral *Lophelia pertusa* (L.) at 64°N on the Norwegian shelf: structure and associated megafauna, Sarsia 80: 145-158.

NPFMC (North Pacific Fishery Management Council). 2000. Draft Environmental Assessment / Regulatory Impact Review for Proposed Amendments 65/65: Harvest Controls for HAPC Biota. North Pacific Fishery Management Council, Anchorage, Alaska, 67 pp.

O'Connell, V.M., W. Wakefield and H. Greene. 1998. The use of a no-take marine reserve in the eastern Gulf of Alaska to protect essential fish habitat. In M. Yoklavich (ed.), Marine harvest refugia for west coast rockfish: a workshop. NOAA Technical Memorandum NMFS-SWFSC-255, pp. 127-134.

Pederson, J. and M. Hall-Arber. 1999. Fish habitat: A focus on New England fishermen's perspectives, American Fisheries Society Symposium 22: 188-211.

Pitcher, R., T. Wassenberg, C. Burridge, G. Smith and R. O'Conner. 1999. The impacts of trawling on large attached tropical seabed fauna, measured in situ. CSIRO Marine Research, Australia. Poster presented at ICES meeting on the Ecosystem Effects of Fishing, March 1999, Montpellier, France.

Risk, M.J., D.E. McAllister and L. Behnken. 1998. Conservation of cold- and warm-water seafans: Threatened ancient gorgonian groves, Sea Wind 10(4): 20-22.

Sainsbury, K.J., R.A. Campbell, R. Linholm and A.W. Whitelaw. 1997. Experimental management of an Australian multispecies fishery: examining the possibility of trawl-induced habitat modification. In E.K. Pikitch, D.D. Hupert and M.P. Sissenwine (eds), Global Trends in Fisheries Management. American Fisheries Society Symposium 20, pp. 107-112.

Smith, J.E., M.J. Risk, H.P. Schwartz and T.A. McConnaughey. 1997. Rapid climate change in the North Atlantic during the Younger Dryas recorded by deep-sea corals, Nature 386: 818-820.

Tilmant, J.T. 1979. Observations on the impact of shrimp roller frame trawls operated over hard bottom communities, Biscayne Bay, Florida. National Park Service, Biscayne National Monument Report Series No. P-553.

Witherell, D., C. Pautzke and D. Fluharty. 2000. An ecosystem-based approach for Alaska groundfish fisheries, ICES Journal of Marine Science 57: 771-777.

van Dolah, R.F., P.H. Wendt and N. Nicholson. 1987. Effects of a research trawl on a hard bottom assemblage of sponges and corals, Fisheries Research 5: 39-54.

THE DISTRIBUTION AND CONSERVATION OF DEEP-WATER CORALS ON CANADA'S WEST COAST

Don E. McAllister[1] and Noel Alfonso[2]*
[1]Canadian Museum of Nature, Ocean Voice International, Ottawa Ontario
[2] Canadian Museum of Nature, Ottawa, Canada

INTRODUCTION

The presence of corals on Canada's west coast is not widely known, nor has their geographic and bathymetric occurrence been well documented. Lack of distributional data is a hindrance to managing resources that benefit from the coral habitat, such as certain rockfish species, and to conserving the biodiversity that throngs in deep-water coral ecosystems.

This lack of data is due, in part, to under-funding of natural history museums, which are seldom able to conduct biological surveys for specimens, and to cutbacks in whole-organism research in universities and basic research in institutions like Fisheries and Oceans Canada (Mosquin et al., 1995). The result has been that non-governmental environmental organizations like the Ecology Action Centre, World Wildlife Fund Canada and Ocean Voice International have been trying to fill the research gap, albeit with slender resources. This paper aims to fill the gap in distributional knowledge using fisheries observer data and museum records.

Concern has risen over the impacts of mobile benthic and other fishing gear on the structure of bottom habitat, the role of essential habitat and keystone species in commercial fisheries, the potential involvement of gear in fisheries declines, and the conservation of biodiversity (Benaka, 1999; Watling and Norse, 1998). Deep-water corals comprise one of the benthic three-dimensional structures that provide shelter, food and other habitat benefits for fishes and invertebrates. These considerations have led to the gathering of preliminary data and initiation of research into deep-water corals off Alaska, Australia, Canada, Iceland, Greenland, Ireland and Norway.

This paper is the first attempt to bring together data on deep-water corals on Canada's west coast for a preliminary assessment of the geographic and bathymetric occurrence of these little-known ecosystems.

ARE DEEP/COOL CORAL STRUCTURES CORAL REEFS?

In this paper we use the term *coral groves* for clusters of coral colonies and avoid, for the moment, applying the term *coral reefs* to British Columbia (B.C.) aggregations. We use the term coral reefs for those densely spaced aggregations of corals that have built up a massive *bioherm* – a carbonate (or occa-

* Please address correspondence to Noel Alfonso, Canadian Museum of Nature, P.O. Box 3443, Station D, Ottawa Ontario K1P 6P4, Canada. E-mail: nalfonso@mus-nature.ca

sionally siliceous) base or platform composed of skeletons of reef-building organisms. Much of the carbonate in tropical coral reefs may originate from calcifying rhodophytes, foraminiferans, molluscs and other organisms, as well as from the symbiotic algal-coral partnership. For further discussion of coral reef definitions see Fagerstrom (1987) and Tomascik et al. (1997: 124-134).

A number of authors further restrict the definition of coral reefs to structures built by hermatypic reef-building organisms – those with symbiotic algae (zooxanthellae) in their tissues – and to those that are wave-resistant, implying occurrence in the well-lit photic and wave-exposed zones. One should note that occurrence of symbiotic algae in cnidarians is not restricted to the tropics. Both zooxanthellae (brown) and zoochlorellae (green) occur on the west coast, e.g., according to Harbo (1999) and O'Clair and O'Clair (1999), in tissues of the giant green anemone, *Anthopleura xanthogrammica*, and the buried green anemone, *Anthopleura artemisia*, but not to our knowledge in any of the corals of the region.

Such restricted definitions (hermatypic and wave-resistant) do have utility in more precisely defining a particular concept of coral reefs and we do not argue against such precision. It can be pointed out that light energy (intensity and seasonal changes in illumination) does, however, decrease with depth and higher latitudes. Wave energy also decreases with depth.

Many of the discussions on coral reef definitions leave out deep-water or cool-water coral assemblages. But coral assemblages called coral reefs, underlain by bioherms, have been reported in Europe (Fosså and Mortensen, 1998) and off Labrador (Smith et al., 1997). These involve dense aggregations of the scleractinian coral *Lophelia*. In the case of the Labrador reefs, the structures are several thousand years old. Clearly such reefs are enduring ecosystems, providing habitats with stability comparable to physical structures. These deep or cool reefs are relatively rich in species compared to other ecosystems at their depths and latitudes. That richness relies, at least in part, on the complex structures and habitats that these coral aggregations provide.

Perhaps the coral assemblages underlain with bioherms at depths or latitudes with light energy insufficient to sustain zooxanthellae could be called *deep* or *cool* coral reefs. Or perhaps a new term should be coined to distinguish these from their tropical counterparts. Further discussion of the possible terminology is probably desirable before a decision is reached.

Whether reef aggregations on the west coast of Canada fit our definition of coral reefs is simply not known. So the term *coral groves* is applied until such time as appropriate investigations (using, for example, submersibles, remotely operated vehicles, sonar, or coring devices) have shown the presence of bioherms under west coast coral aggregations.

SITES, MATERIALS AND METHODS

Data include 32 gorgonian records from the Royal British Columbia Museum, Victoria, B.C., 353 records of deep-water corals from the Department of Fisheries and Oceans (DFO) fisheries observers program, and eight gorgon-

List of Coral Species Known from Canada's West Coast

Phylum Cnidaria
Class Anthozoa
Subclass Octocorallia (=Alcyonaria) - octocorals
Order Gorgonacea - sea fans and sea whips
Acanthogorgiidae
Calcigorgia spicculifera Broch, 1935
Briareidae
Anthothela pacifica (Kükenthal, 1913)
Chrysogorgiidae
Chrysogorgia sp.
Isididae
Isadella sp.
Paragorgiidae
Paragorgia pacifica Verill 1922 (=*P. arborea auctorum*)
Plexauridae
Eunicea sp.
Psammogorgia spauldingi (Nutting, 1909)
Swiftia torreyi (Nutting, 1909)
Primnoidae
Callogorgia kinoshitae Kükenthal, 1913
Parastenella sp. aff. *Spinosa* (Wright & Studer, 1889)
Primnoa willeyi Hickson, 1913
(=*P. reseda* & *P. resedaeformis auctorum*)
Order Alcyonacea - soft corals
Alcyoniidae
Anthomastus sp. aff. *grandiflorus* Verill, 1878
Clavulariidae
Clavularia spp.
Nephtheidae
Gersemia rubiformis (Ehrenberg, 1834)
Subclass Hexacorallia (=Zooantharia)
Order Scleractinia (= Madreporaria) – hard, stony or 'true' corals
Caryophylliidae
Cyathoceras quaylei Durham, 1947
Desmophyllum cristagalli Milne Edwards & Haime, 1849
Paracyathus stearni Verill, 1869
Solenosmilia variabilis Duncan, 1873
Dendrophylliidae
Balanophyllia elegans Verill, 1864
Subclass Ceriantipatharia
Order Antipatharia - black or thorny corals
Heterotainiales
Antipathes sp.
Bathypathes patula Brock, 1889

ian records from the collections of the Canadian Museum of Nature, Ottawa, for a total of 393 records.

The majority of records, those from DFO fishery observers, were identified to order (usually the scientific name, but in some cases as "unident. stony corals," Scleractinia). Of the museum specimens, 34 were identified to genus, 27 to species (all Gorgonacea). The lack of taxonomic revisions and identification guides hampers the identification of many groups of invertebrates in and off Canadian territorial waters.

The classification of the orders of corals in our west coast data set is outlined above. Information down to the species level is provided for the Gorgonacea. The classification follows Brusca and Brusca (1990) and Austin (1985), who supplied most of the records for B.C. and the species names. Harbs (1999) illustrates some of the species in colour.

There are 21 species in this assuredly incomplete list.* That compares with 31 species known from the Gulf of St. Lawrence and off Nova Scotia on the Atlantic coast of Canada (Breeze et al., 1997; Brunel et al., 1998).

Distribution data are biased by the fact that much of the data was obtained from fish catch sampling, so samples tend to be drawn from fishing grounds and from depths in which trawls operate. On the other hand fisheries are fairly widespread in British Columbia, so the general pattern shown is valid. It is likely that corals do occur in the gaps between recorded locations in many of the areas not covered by the observer program. It is also possible, as some maintain, that areas rich in corals are attractive to fishes so that fishers set trawls on or near coral forests (Breeze et al., 1997), despite a tendency for rougher bottom. It should be emphasized that fishers aim to catch fishes, not harm habitat.

Another source of potential bias is possible clearance of corals by fishing gear prior to the 1997 sampling year. McAllister (1999) indicated that trawling is moving into deeper waters as shallow-water stocks of fishes are reduced and as more powerful deep-water-capable fishing gear is developed. It is possible that coral colonies have already been reduced in the 20 to 200 m range. We provide some information from longline shrimp trap fishing that supports this hypothesis. An analysis of older museum records and biological surveys might provide data to further test this hypothesis.

Data from divers, mostly from published sources, are presented in Table 2. Scuba divers' data originate closer to shore and tend to be in shallower water than trawl data.

RESULTS

The Canadian records in our data set range from Dixon Entrance north of Haida Gwaii (Queen Charlotte Islands) in the north to southern Vancouver Island in the south. Inshore to offshore the corals range from the inner waters of Hecate and Georgia Straits, including fjords, to the offing of Dixon Entrance, Haida Gwaii and Vancouver Island. In addition there are a few records from the Gulf of Alaska and off Washington State, USA (only some of the latter are mapped, see Figure 6). Distribution of the orders and the few specimens identified to species are shown in the maps, Figures 1 to 6.

Recorded depths ranged from 20 to 2057 m, with most lying between 100 and 500 m. Table 1 shows the depth distribution of 375 west coast coral samples. Omitted from the table are 17 records lacking depth data and one specimen identified only to subclass.

* The list excludes hydrocorals and cupcorals known from B.C.

Figure 1. Distribution of sea fans (Gorgonacea, Octocorallia) on the west coast of Canada

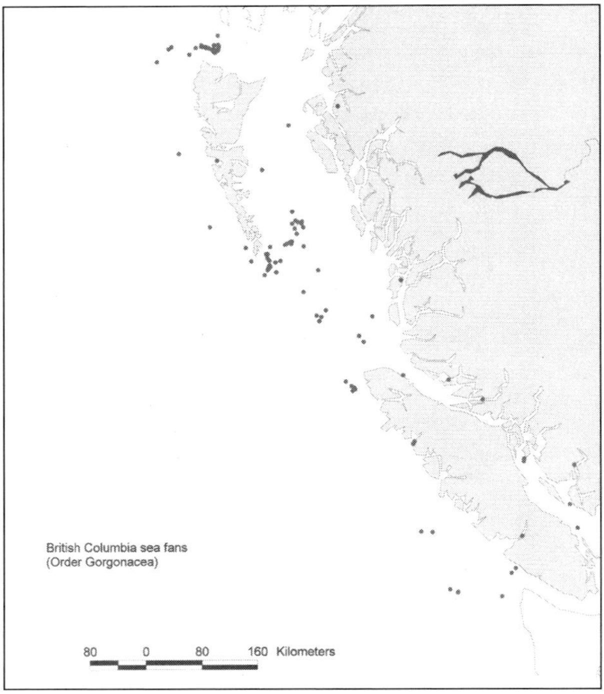

Figure 2. Distribution of soft corals (Alyconacea) on the west coast of Canada

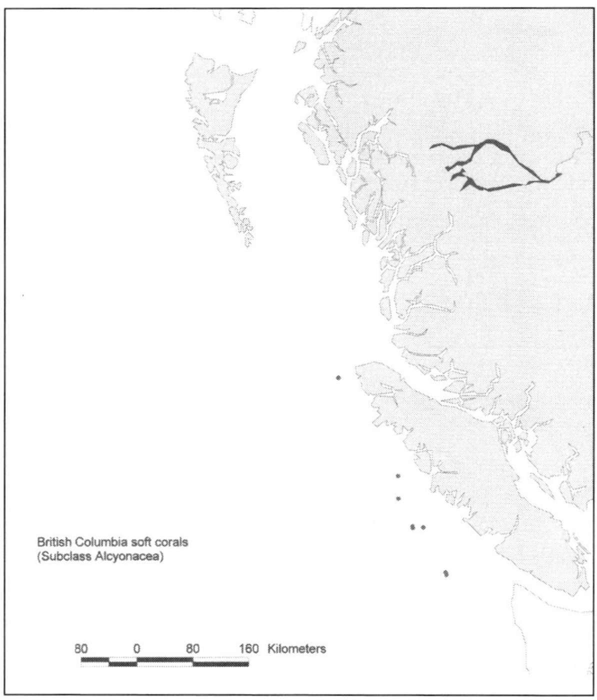

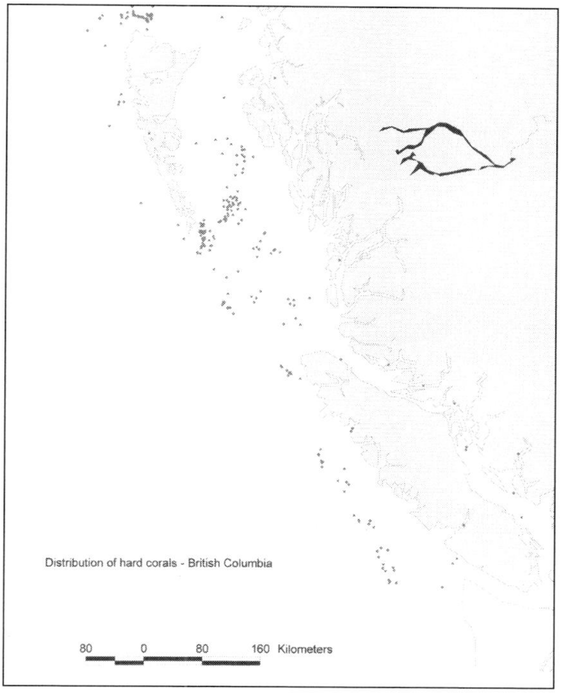

Figure 3. Distribution of hard or stony corals (Scleractinia [=Madreporaria]) on the west coast of Canada

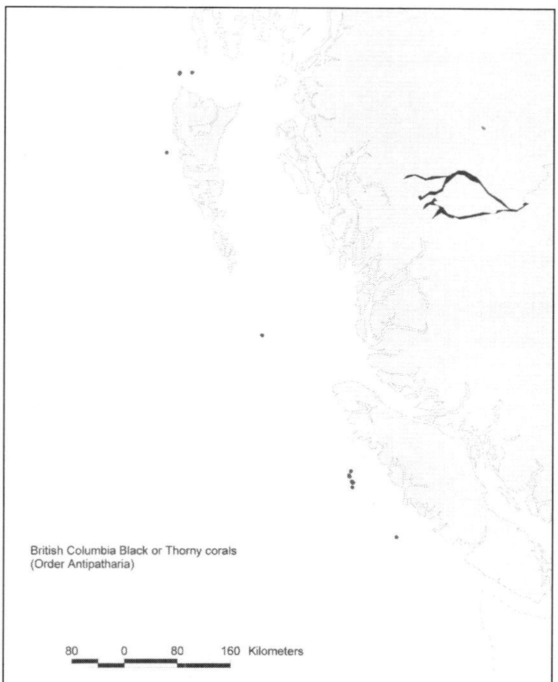

Figure 4. Distribution of black corals (Antipatharia) on the west coast of Canada

Figure 5. Distribution of Paragorgia pacifica, *Paragorgiidae, on the west coast of Canada*

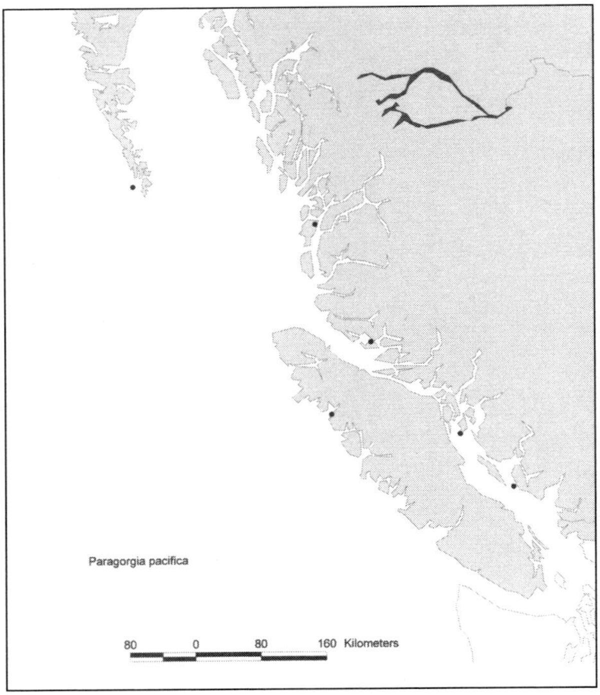

Figure 6. Distribution of Primnoa willeyi, *Primnoidae, on the west coast of Canada and Alaska*

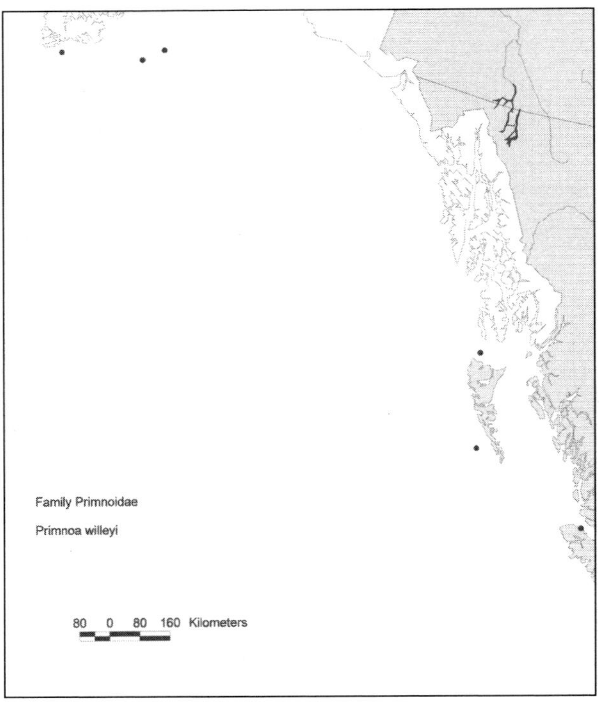

The sampling method does not provide information about the size of coral groves. They could be: (i) small patches or individual colonies, isolated by considerable distances, (ii) small patches clustered relatively closely, (iii) large patches of considerable extent, or (iv) some mixture of these patterns. One Norwegian *Lophelia* coral grove is reported to be 15 km long. The size and proximity of coral groves is of interest in terms of both fisheries and conservation. It would be desirable to put some effort into using submersibles or ROVs to find out what are the most common patterns. Scuba observations suggest that groves are patchy, but most of the groves are beyond easy diving depths and the modal depths of even the soft and hard corals, let alone the black corals.

Table 1 also compares depth frequency data for the four groups of corals. About half of the records of Gorgonacea (51%) and Scleractinia (49%) occur on the continental shelf at depths of between 20 and 200 m. Most Alcyonacea (67%) and Antipatharia (89%) were collected deeper than the continental shelf, at between 401 and 600 m. However, the number of Alcyonacea is too low to permit much confidence in this indication of depth occurrence. This interpretation of depth occurrence is constrained by the fact that most of the samples were drawn from fishing locations with unequal sampling of depth zones.

A chi-square test was carried out on the depth distribution differences between the Gorgonacea, Scleractinia and Antipatharia; Alyconacea coral records were regarded as too few to be analyzed. Depth categories were grouped into 20-200 m, 201-400 m, and 401- 600+ m. A test between depth frequencies of Gorgonacea and Scleractinia gave a chi-square value of 0.30 with 2 d.f., and $p = 0.9\text{-}0.8$. The maps suggest that many gorgonacean stations coincide with or are proximate to scleractinian stations. Since the differences were not significant the two groups were lumped and tested against depth frequencies of Antipatharia. The latter test yielded a chi-square of 85 with 2 d.f. and $p< 0.001$. That suggests that Antipatharia tend to inhabit significantly deeper water than do Gorgonacea and Scleractinia. But a larger sample of Antipatharia would be desirable to better test that finding.

Table 1. Depth ranges of west coast Canadian corals number and per cent of stations.

Depth range (m)	Gorgonacea - sea fans	Alcyonacea - soft corals	Scleractinia - stony corals	Antipatharia - black corals
20-100	19 (12%)	1	22 (12%)	0 (0%)
101-200	62 (40%)	2	71 (37%)	2 (11%)
201-300	57 (36%)	0	27 (14%)	2 (11%)
301-400	5 (3%)	1	53 (28%)	0 (0%)
401-500	1 (1%)	1	7 (4%)	10 (56%)
501-600	11 (7%)	4	11 (6%)	4 (22%)
1,485 & 2,057	2 (1%)	0	0 (0%)	0 (0%)
TOTAL	157	9	191	18

Gorgonians tend to live on vertical or sloping rock walls, ledges, rocky outcrops and boulders. In B.C., divers' observations show corals attach themselves to the vertical walls of the U-shaped, glacier-carved fjords (pers. com. Dr. Mark Graham). The distribution maps for Gorgonacea and Scleractinia suggest that these groups occur in the outer half of fjords, such as Alberni, Jervis and Knight inlets. Restriction to the outer half of the fjord might be related to the higher freshwater or silt levels or lower plankton levels in the inner portion. The bases are normally attached to a hard surface. In addition they prefer locations where currents sweep past the sea fans, carrying plankton to the polyps. Those gorgonians which grow in a single plane, tend to have that plane at right angles to the current. That would ensure that the polyps on most branches will have more equal access to plankton.

Figure 7 shows a map of the bottom-fish trawling grounds off the west coast of Canada, based on information provided by Fisheries and Oceans Canada. The degree of coincidence between the trawling and the deep-water coral data should not be overemphasized, given that the museum specimens are not numerous enough to provide a control set of data. One can surmise, however, that where benthic mobile fishing gear, or gear that is likely to catch on corals, is used, then it is likely that there will be fishing impacts on coral ecosystems. The occurrence of corals as bycatch in over 350 trawl stations does not negate that hypothesis.

The occurrence of corals in otter trawls used for fishes (Behnken, 1993a, 1993b) and direct submersible observation of otter trawl tracks in Alaska (Krieger, 1998) show that otter trawls do damage corals in the eastern North Pacific. Krieger's (1998) study also showed that about 50% of the coral in the trawl path was removed or broken by the trawl gear, and that impacts were still visible 17 years later though broken coral was partially covered with living tissue.

Shrimp beam trawling gear used in B.C., according to Deb Logan (e-mail, 12 September 1997) and our own observations (D.E.M.) in southern and northern B.C. as well as in southern Alaska north to Prince William Sound, suggest that shrimp beam trawls do not put corals at risk. This may reflect the preferred type of bottom (mud or sand) this gear is used on or to the greater buoyancy of these beam trawls compared to otter trawls. Beam trawls also lack the heavy, usually metal doors that are used in bottom otter trawls. Midwater trawls are typically fished in the water column above the bottom and do not impact corals unless they accidentally touch the bottom.

Commercial longline prawn trap gear, according to Deb Logan (op. cit.), commonly caught tree coral in B.C., where her family fished, for several years starting in 1974. But by 1980, little coral was caught by the gear. In 1997 other prawn trap fishers reported to her that little if any coral was being caught in the same area. These data indicate that longline prawn trap gear can catch and diminish coral colony numbers. Further the data indicate that present-day absence or near-absence of coral colonies cannot be taken as indicating original conditions when certain gear types have been actively used.

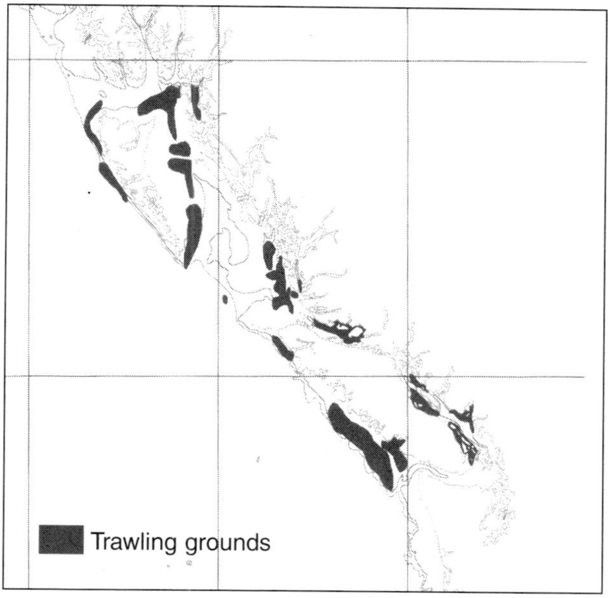

Figure 7. Trawling grounds for fishes and shrimps based on information from Fisheries and Oceans Canada

Aside from fishing gear there are other potential threats to deep-water corals. Dumping from land-based sources might cover coral colonies or result in turbidity that would interfere with feeding activity. Larval life and physiology are too poorly investigated to know whether surface or benthic stages might suffer from oil spills resulting from tanker traffic, ship wrecks or ballast-water dumping. Similarly it is not known whether persistent organic chemicals might affect growth or survival of corals. It is possible that climate warming might reduce productivity through stabilization of the water column, and by reducing the supply of nutrients in the photic zone and the plankton food supply, though there is now sufficient evidence to show that surface waters off B.C. are warming.

A number of other species use deep-water corals as habitat. Analysis of the fisheries observer data for catch sizes in hauls with corals and without corals would permit quantitative analyses to be carried out for commercial fishes.

Literature data for Alaska and British Columbia (Risk et al., 1997) provide evidence that the following vertebrates and invertebrates are using coral habitat (see Figure 8 for examples): fishes include decorated warbonnets (*Chirolophis decoratus*), rougheye rockfish (*Sebastes aleutianus*), shortraker rockfish (*Sebastes borealis*), black rockfish (*Sebastes melanops*), yelloweye rockfish (*Sebastes ruberrimus*) and china rockfish (*Sebastes nebulosus*); the invertebrates include sponges, hydroids, hydrocorals, anemones, scallops, octopuses, shrimps, bryozoans and basketstars. Juveniles of the basketstar, *Gorgonocephalus eucnemis*, are often found clinging to branches of the soft coral *Gersemia rubiformis* (Harbo, 1999).

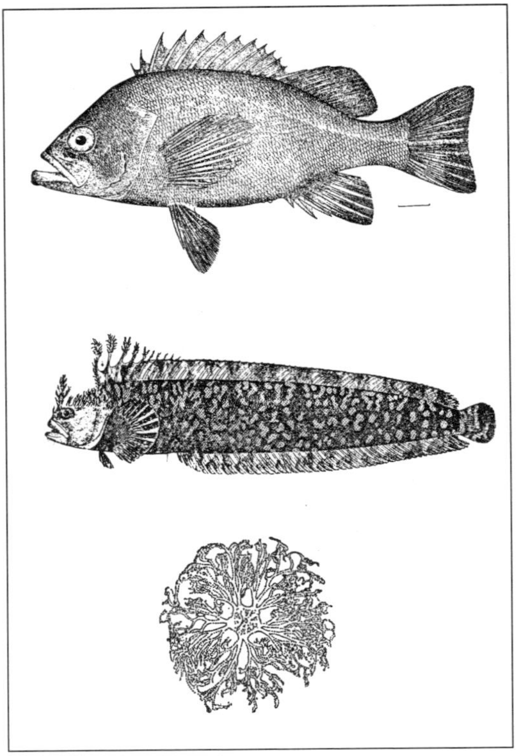

Figure. 8. Black rockfish, Sebastes melanops *(above), decorated warbonnet,* Chirolophis decoratus *(middle), and basketstar,* Gorgonocephalus, *three of the many inhabitants of west coast deep-water coral groves.*

The rockfishes are commercially fished. B.C. catches were 24,456 metric tonnes in 1989 and 14,903 tonnes in 1994 (Statistics Canada, 1997; Fisheries and Oceans Canada website). In addition rockfishes are the basis of sustenance and recreational fisheries by First Nations, local communities and sports fishers. Management of these rockfishes, like that of deep-water corals, is challenged by their slow growth rates, with ages reaching several decades.

Divers' observations, magazines, books and guides provide information on corals at the shallower end of their depth range. Table 2 summarizes some of the information from these sources (see box page 137). As well those publications are a source of colour photos of corals in their natural habitat. The number of publications attests to the growing dive industry and interest in sea life. Marine protected areas foster this interest.

Pharmaceutical companies have shown considerable interest in the organic compounds, such as prostaglandins and antibiotics, found in tropical gorgonians (Risk et al., 1997). Deep-water species may also contain compounds of interest, providing an additional incentive for their conservation.

We lack age data for west coast Canadian corals. However, *Primnoa reseda* (related to west coast *Primnoa willeyi*) from Nova Scotia has been aged at

Table 2. Diver records of coral occurrence in British Columbia

Geographic location and depth	Sources and notes
Queen Charlotte Strait near Port Hardy as shallow as 24 m	Gordon Heath, e-mail, 24 June 1997. A small species about 30 cm max height, darker pink to white.
Queen Charlotte Strait in Ripple Passage near Barry Islet in 30 m	Britnell (1997). An expansive garden of tiny gorgonian corals on a rock wall plus a deeper slope festooned with opulent fields of orange and white gorgonian corals.
Queen Charlotte Strait on the west side of Browning Passage 3-30+ m depth	Britnell (1997). Soft corals on a steep wall drop-off. Swanson (1993). Pink soft coral in clumps 30 cm across.
Discovery Passage between the city of Campbell River and Quadra Island	Pemberton (2000b). Profusion of yellow cup corals (presumably *Balanophylla elegans*, family Caryophyllidae).
N. Agamemnon Pass at the entrance to Sechelt Inlet and at the Fearnie Bluffs Georgia Strait S. entrance to Agamemnon Pass	Pemberton (2000a), Pratt-Johnson (1995). One of the very few places in B.C. where divers can find large fans of the rare crimson gorgonian coral – located on a steep wall dive in excess of 150 feet (46 m).
Grant Island, Welcome Passage	Pratt-Johnson (1995). Bright red dwarf gorgonian coral and hard pink hydrocoral.
Strait of Georgia, off Sechelt Peninsula, starting as shallow as 49 m	Gordon Heath, e-mail, 24 June 1997. Large sea fans.

A diver's view of coral grove biodiversity.

Off a small islet in Queen Charlotte Strait, we descend past a wall covered in delicate brooding anemones towards a sight I have heard about for several years but never seen. At a depth of twenty-five metres, we suddenly enter a small undersea canyon, and I catch my breath as I take in the wondrous sight that surrounds me: an entire field of gorgonian corals in various shades of pink and white, their delicate fans swaying gently in the current. Moving close to photograph several of the twenty-centimetre-high fans, I disturb a swimming scallop. It lifts off the bottom and darts away from me, resembling a swimming set of false teeth as it bites its way through the water. One of my companions waves me over to where a giant Pacific octopus slowly winds its way through the maze of coral... Suddenly, from underneath the outstretched tentacles of the octopus, a magnificent Decorated Warbonnet emerges, a fitting final image of a truly magical place.

Swanson (1993)

about 300 years old (Heikoop et al., 1998) using radioactive signatures. Scleractinian coral reefs (the built-up structure of dead and living corals) off Labrador were several thousand years old and individual specimens are more than 100 years old, according to Smith et al. (1997).

Deep-water west coast Canadian coral ecosystems are widespread and used by numerous species. This suggests that these ecosystems are a valuable resource for fisheries and biodiversity. The elevated age that many corals attain suggests that these coral grove structures will take very long periods of time to be replaced if the forces that degrade them are discontinued. These elements suggest that deep-water west coast coral groves are deserving of conservation. New resources should be secured to provide more detailed information on their location, ecology and taxonomy. In the meantime, efforts should be made to include a diversity of coral groves, at different depths, in marine protected areas that are free from mobile fishing gear or other activities that might destroy them.

BENEFITS OF KEEPING/COSTS OF LOSING DEEP-WATER CORALS

Benthic trawling and other human activities affect deep-water coral ecosystems. Fishing on coral groves is initially attractive because of the associated fish congregations. But fishing activities that harm corals will result in long-term costs.

Some of the benefits of keeping and costs of losing deep-water coral groves are:

- Corals provide habitat for commercial species, enhancing their populations and resultant catches. Loss of essential fish habitat will result in reduced catches.
- Deep-water corals provide key habitats for a proportion of the west coast's rich biodiversity. Loss of this habitat is likely to result in a reduction in biodiversity.
- Deep-water corals such as gorgonians possess unique chemical compounds, promising to pharmaceutical and other industries. Conserving corals means keeping such options open.
- Deep-water corals and their sub-fossils contain records of past climates. Ensuring survival of old-growth corals means guarding this climate database and perhaps other opportunities, for future research and management.
- Coral groves attract resident and visiting scuba divers and submersible tours and help promote a vibrant source of employment and profits.
- Corals possess their own intrinsic value.

OPTIONS FOR SAFEGUARDING DEEP-WATER CORALS

The benefits just discussed make a case for safeguarding deep-water corals from harmful human activities. Here we discuss some key options that might be useful in conserving corals.

1. *Fishery reserves and protected areas.* Setting aside healthy stands of corals or degraded stands for restoration would result in enhanced commercial catches of fishes and conservation of biodiversity, as well as ancillary spinoffs. Fishers, conservationists, park biologists/ planners, fishery biologists, environmentalists and divers can work together to set aside important stands of corals, protecting them from activities that break off or smother the colonies. In Norway the fishers' council saw advantage in protecting corals as essential fish habitat for fish population enhancement. Coral groves were set aside as trawl-free areas. Canada has the capacity to establish marine protected areas under legislation administered by Parks Canada, Fisheries and Oceans Canada, and Environment Canada. As yet no marine protected area has been established on the west coast to protect corals for the biodiversity they harbour or fish stocks they enhance.
2. *Low habitat impact fishing gear.* Switching to or developing fishing gear with low habitat impact provides a second option. The amount of disturbance of bottom habitat varies with the type of fishing gear. As the benefits from such changes extend beyond the fishing industry, governments should consider supporting gear conversion and enhancement. Fishing moratoria or periods of time when stocks are depressed provide optimal opportunities for gear conversion. Industry and fisheries management have already agreed in principle, e.g., Principle 2.3 of the *Canadian Code of Conduct for Responsible Fishing Operations* states, "Ensure that fisheries activities are not conducted in a fashion that would endanger fish stocks or the environment" (DFO, 1998). Similarly the FAO Code (1995), which Canada has agreed to, calls upon participants to provide that, "7.2.2. f) adverse environmental impacts on the resources from human activities are assessed and, where appropriate, corrected." Canada also has the capacity to undertake fish habitat protection under the *Fisheries Act* and has undertaken responsibilities under the Biodiversity Convention to conserve biodiversity and to use biological resources sustainably.
3. *Research on deep-water corals.* Taxonomic and ecological research on corals in Canada lags behind that in Europe and Australia. We have only just begun mapping the locations where corals occur. Detailed hierarchical marine ecosystem classification systems have only recently been established on Canada's east and west coasts (Day and Roff, 2000; McAllister, 2000). Biological surveys would provide specimens for taxonomic research, the identities of species inhabiting coral groves, and data for mapping. Basic studies on coral groves on the west coast are needed to determine:
    - How large an area do coral groves occupy?
    - What are the average and maximum heights of coral colonies?
    - How old are the colonies?
    - Do west coast corals contain climate record data?

- What are the usual substrates?
- Are groves underlain by bioherms? Do coral aggregations comprise coral reefs?
- When does coral spawning occur? Settlement of larvae?
- What are the usual current speeds in coral grove locations?
- What proportion of coral groves show signs of trawl activities? How much grove area is affected?
- When do polyps open? What do they feed on?
- Which non-coral species inhabit coral groves? In what numbers and sizes? Degree of occurrence inside and outside of coral groves?
- What are the preferred temperatures, salinities and depths of each coral species?

We also need to:

- settle the taxonomic relationships and names of corals;
- provide easy-to-use identification guides;
- map the occurrence of coral species; and
- educate the public about deep-water corals.

The co-operation of government departments, natural history museums, universities, public aquaria, and environmental organizations would help reach those goals. Environmental organizations have made a good start in raising public awareness and in gathering existing information from fishers, museums and the literature. Now co-operative follow-through is needed.

Tropical coral reefs have been, along with tropical rainforests, the centre of attention of extensive programs for conservation and sustainable use. It is time that we turned our attention to the equally fascinating but virtually unknown coral groves of the deeper and cooler waters of the planet.

CONCLUSIONS

We conclude that:

1. Corals occupy significant parts of the continental shelf and slope of the west coast of Canada.
2. Gorgonacea are most common between 101 and 300 m depth, Scleractinia between 101 and 400 m depth, and Antipatharia between 401 and 600 m depth. Distributional coherence between Gorgonacea and Scleractinia suggests that the same marine protected areas could function to conserve both, while Antipatharia would require separate and deeper marine protected areas.
3. Available evidence indicates that these coral groves provide important habitat to commercial and non-commercial species.
4. Our data show that benthic trawling gear does impact corals, by breaking off portions of colonies *in situ* and removing some of the fragments as trawl bycatch.
5. That west coast coral groves contain centuries-old individuals is suggested by evidence from the Atlantic coast. That in turn suggests that

protecting these ecosystems is to be preferred to trying to restore them after they have been lost.

6. Benefits of coral groves include provision of essential habitat for commercial species (biological resources), biodiversity and potential pharmaceutical producers; records of past climates; and an attraction for the growing underwater recreational dive industry. Loss of stock enhancement and other benefits means lowered catches, lost employment for fishers and others, and reduced profits for the fishing industry. Failure to protect shallower near-shore coral groves would affect the dive industry.
7. Options for sustaining the benefits of deep-water coral ecosystems include the establishment of no-take marine protected areas including fisheries reserves; switching to or developing more habitat-friendly fishing gear; and conducting research to provide basic biological information so as to better understand, conserve and manage these corals.
8. Given the accumulating damage to coral groves and the loss of benefits, high priority should be given to identifying candidate areas and establishing marine protected areas, free from human disturbance, for deep-water coral ecosystems.
9. Research on ecology, taxonomy and distribution of deep-water corals is needed to maximize conservation and resource management goals. If you can't name or know their habitat or geographic distribution, it is difficult to conserve and manage these species. The taxonomic and ecological impediments to this vital knowledge call for new research and investment in marine organisms.
10. The fisheries observer program is a valuable source of data thanks to the support of fishers. The data could be made more valuable if a modest amount of funding was provided to natural history museums to deposit, identify to species, and catalogue representative voucher specimens from different geographic areas, depths and seasons. The contribution of fishers deserves recognition.
11. International and national programs have been developed to conserve and sustainably use tropical coral reefs. It is time to develop similar programs with the same goals for deep-water coral reefs.

ACKNOWLEDGEMENTS

We thank Lisa A. Kirkendale (former fisheries observer on trawlers, but then working for the Royal British Columbia Museum) for compiling the data from Fisheries and Oceans Canada and the Royal British Columbia Museum in a database. We further express our gratitude to Fisheries and Oceans Canada staff Barry Ackerman (Groundfish Trawl Co-ordinator) and Rick Stanley (Fisheries Dynamics Division), and Ken Derksen, Archipelago Marine Research Ltd., for fisheries observer records; and Laura Richard, A/Regional Director, Sidney, for trawling ground information. Above all we thank the fishers who helped provide key data on the deep-water corals, and paid for and supported the observer program. Those data are vital for sustainable fisheries management and conservation of marine biodiversity. Dr. Philip Lambert, Invertebrate Curator, and Mr. Kelly Sendall, Invertebrate Technician of the Royal British Columbia Museum, provided collection data under their care. Dr. Jean-Marc Gagnon, Chief Collection Manager at the Canadian Museum of Nature, permitted use of collection data under the museum's care.

Gordon Heath, diver and board member of the Marine Life Sanctuaries Society, generously provided records of corals seen while diving.

Ocean Voice International thanks ESRI headquarters in Redland, California, who provided them with the ArcView GIS software, and the British Oceanographic Data Centre, who provided its digital GEBCO marine chart set of the world at a reduced price.

Parks Canada generously provided funds that enabled the database to be converted to ArcView format and mapped.

REFERENCES

Austin, W.C. 1985. An annotated checklist of marine invertebrates in the cold temperate northeast Pacific. Khoyatan Marine Laboratory, Cowichan Bay, B.C, Canada. Vol. 1, 218 pp.

Benaka, L.R. 1999. Fish habitat: Essential fish habitat and rehabilitation. Proceedings of the Sea Grant Symposium on Fish Habitat: "Essential Fish Habitat" and Rehabilitation, Hartford, Connecticut, U.S.A., 26-27 August 1998. American Fisheries Society, Bethesda, Maryland, 459 pp.

Behnken, L. 1993a. Southeast Alaska trawl closure: A case study in risk-averse management, Sea Wind 7(1): 8-14.

Behnken, L. 1993b. Photos of southeast Alaska fishing bank habitat sensitive to trawl damage, Sea Wind 7(2): 25-28.

Breeze, H., D.S. Davis, M. Butler and V. Kostylev. 1997. Distribution and status of deep-sea corals off Nova Scotia. Marine Issues Committee Special Publication Number 1. Ecology Action Centre, Halifax, N.S., 58 pp.

Brunel, P., L. Bossé and G. Lamarche. 1998. Catalogue of the marine invertebrates of the estuary and Gulf of St. Lawrence. Canadian Special Publication of Fisheries and Aquatic Sciences 126: 1-405.

Britnell, J. 1997. In quest of adventure. Exploring the majesty of British Columbia's Queen Charlotte Strait, Diver, Richmond, B.C., March 2000: 4-6.

Brusca, R.C. and G.J. Brusca. 1990. Invertebrates. Sinauer Associates, Inc., Sunderland, Mass., 922 pp.

Day, J.C. and J. Roff. 2000. Planning for representative marine protected areas. A framework for Canada's oceans. Report prepared for World Wildlife Fund Canada, 147 pp.

DFO. 1998. Canadian code of conduct for responsible fishing operations. Secretariat of Fisheries and Oceans Canada, Ottawa, 24 pp.

Fagerstrom, J.A. 1987. The evolution of reef communities. John Wiley and Sons, New York, 600 pp.

FAO. 1995. Code of conduct for responsible fisheries. Food and Agricultural Organisation of the United Nations., Rome, 41 pp.

Fosså, J.H. and P.B. Mortensen. 1998. Artsmangfoldet på *Lophelia*-korallrev og metoder for kartlegging of overvåkning, Fisken og Havet Number 17: 1-95.

Harbo, R.M. 1999. Whelks to whales. Coastal marine animals of the Pacific Northwest. Harbour Publishing, Madeira Park, B.C., 245 pp.

Heikoop, J.M., M.J. Risk, A.V. Lazier and H.P. Schwarcz, H.P. 1998. δ^{18} O and δ^{13} C signatures of a deep-sea gorgonian coral from the Atlantic coast of Canada. Spring American Geophysical Union, Boston Meeting. Abstract.

Krieger, K. 1998. *Primnoa* spp. observed inside and outside a bottom trawl path from a submersible. Abstract. 10th Western Groundfish Conference, p. 88.

McAllister, D.E. 1999. A global trawling ground survey. Report for World Resources Institute, Marine Biological Conservation Institute and Ocean Voice International, 25 pp. + Appendices A to E.

McAllister, D.E. 2000. A model framework for a marine ecosystem classification for Canada's West Coast. Report prepared by Ocean Voice International for Parks Canada, 89 pp.

Mosquin, T., P.G. Whiting and D.E. McAllister. 1995. Canada's biodiversity. The variety of life, its status, economic benefits, conservation costs and unmet needs. Canadian Museum of Nature, Ottawa, 293 pp.

O'Clair, R.M and C.E. O'Clair. 1999. Southeast Alaska's rocky shores. Animals. Plant Press, Auke Bay, Alaska, 564 pp.

Pemberton, D. 2000a. Diving the end of Highway 101, Diver, Richmond, B.C., February 2000: 34-35.

Pemberton, D. 2000b. In the depths of Discovery Passage, Diver, Richmond, B.C., April 2000: 36-37

Pratt-Johnson, B. 1995. 101 dives from the mainland of Washington and British Columbia. Heritage House Publishing Co. Ltd., Surrey, B.C., 320 pp.

Risk, M.J., D.E. McAllister and L. Behnken. 1997. Conservation of cold- and warm-water sea fans: Threatened ancient gorgonian groves, Sea Wind 12(1): 2-21.

Smith, J.E., M.J. Risk, H.P. Schwarcz and T.A. McConnaughey. 1997. Rapid climate change in the North Atlantic during the Younger Dryas recorded by deep-sea corals, Nature 386: 818-820.

Statistics Canada. 1997. Canada yearbook 1997. Statistics Canada, Ottawa, 515 pp.

Swanson, D. with photos by D. Sanders. 1993. The emerald sea. Exploring the wilderness of the North Pacific coast. Whitecap Books, Vancouver, 147 pp.

Tomascik, T., A.J. Mah, A. Nontji and M.K. Moosa. 1997. The ecology of Indonesian seas. Periplus Editions (HK), Ltd., 642 pp.

Watling, L. and E.A. Norse (section eds). 1998. Special section: Effects of mobile fishing gear on marine benthos, Conservation Biology 12(6): 1178-1240.

MARINE PROTECTED AREA FRAMEWORK FOR DEEP-SEA CORAL CONSERVATION

Christopher A. Miller
Wetlands Research Centre and Department of Biology
University of Waterloo, Waterloo, Ontario, Canada

SUMMARY

Systems of marine protected areas can be used to safeguard deep-sea coral communities from anthropogenic activities that threaten their ecological integrity. Bottom trawling and mineral exploration are of particular concern in this regard. Such reserve systems, however, should be designed using a systematic, science-based approach, instead of an opportunistic philosophy that may leave ecologically significant areas unprotected. Strategies involving representivity, hotspots, rarity, productivity and focal species are all applicable to a spaces-orientated approach to deep-sea coral conservation. A representative approach would seek to protect features typical of a given biogeographical region of which corals may be an intricate component. Although all deep-sea coral areas are arguably ecological hotspots, those assemblages that are oldest, largest, most biologically diverse and least fragmented should probably be given the highest priority for protection. A rarity approach should give conservation priority not only to areas with, or thought to contain, rare deep-sea coral species, but also areas with rare species dependent upon coral as habitat. Deep-sea corals sometimes occur where productivity rates are naturally high. Thus, by securing from development the most productive areas of the ocean, deep-sea corals may be protected as byproducts. Although all of these approaches can be used to select coral areas in greatest need of protection, they can also be used to justify why a known occurrence should be protected. Regardless of which methods are utilized, the precautionary principle and the concept of redundancy should be included in any system of marine protected areas that targets deep-sea corals in order to sufficiently address the biodiversity crisis. Nova Scotia's deep-sea corals need to be protected within the marine reserves of Parks Canada and the Department of Fisheries and Oceans using the above approaches. Interim conservation measures of the Fisheries Act, in the form of gear restrictions and temporary closures, can be used to protect candidate coral sites until such time as they can be formally secured from development through legislation.

INTRODUCTION

Given their slow-growing physiology and sensitivity to disturbances, deep-sea corals are intrinsically vulnerable to bottom trawling. Such a fishing technique can reduce the structural diversity of the ocean floor, alter the biogeochemical cycles of the water column, and expose, remove, crush and bury benthic communities (Watling and Norse, 1998). A controlled study off the central coast of California demonstrated that bottom trawling with mobile fishing gear can reduce the habitat complexity of the ocean floor and enhance the abundance of opportunistic species that prey upon the diet of commercial groundfish (Engel and Kvitek, 1998). Areas that contain deep-sea corals may be particularly vulnerable to such impacts because they typically occur on the continental shelf and slope where few natural disturbance regimes are present, thus giving them a predisposition to anthropogenic degradation. Oil and gas exploration can also inflict damage by introducing amplified sediment loads into coral areas or by physically crushing corals through drilling, pipe-laying and exploration activities. In order to protect deep-sea corals from these types of human disturbances, there is a need to adopt a spaces-orientated approach to conservation and to establish systems of marine protected areas that exclude activities that threaten the ecological integrity of coral areas. Mills and Carlton (1998) point out that marine reserves are greatly needed in the offshore, even in international waters beyond a country's 200-mile limit. Although other conservation strategies involving gear restrictions and temporary closures may help protect deep-sea corals in the short term, marine protected areas offer a more permanent solution. It is important to select, therefore, the coral areas in greatest need of protection using a systematic, science-based approach to protected areas design.

The science of reserve design is well developed in the terrestrial realm, but remains inadequate in the marine environment, perhaps because systems of marine protected areas are fundamentally more difficult to design. Transboundary issues are more pronounced in the marine environment than on land. Pelagic species move widely in, and out of, marine protected area boundaries, and benthic species located within a marine reserve may be linked to nutrient sources many kilometres away. Marine protected areas are also structurally more complex than their terrestrial counterparts because they often contain a third dimension that terrestrial protected areas lack. It is often challenging to acquire information about the ocean floor, thus making it hard to determine where individual protected areas should be located in order to maximize attained conservation benefits. Since marine habitat conversions are less visible to the general public than terrestrial disturbances, fewer people are aware of the degree to which the continental shelves have been domesticated. As a result, significant areas of the ocean remain unprotected because a strong public voice, opposed to the loss of marine habitat, is often lacking.

The complexity associated with the design of marine protected areas is often used as an excuse for inaction, and an opportunistic approach is sometimes adopted instead. Such a philosophy, however, may encourage marine protected areas to be established where they are easiest to implement rather than where they are most needed, thus leaving ecologically significant areas, such as deep-sea coral assemblages, unprotected and vulnerable to degradation. Deep-sea corals can serve as a catalyst to shift marine protected area thought from such an opportunistic approach to a more ecologically meaningful one. On land, North America's first terrestrial protected areas were primarily established to conserve awe-inspiring landscapes and aesthetically pleasing vistas. They include places such as Banff and Cape Breton Highlands National Parks in Canada, and the wilderness areas of Yellowstone, Yosemite and Grand Canyon National Parks in the United States. Not until the Everglades National Park was implemented in 1934 to protect a unique ecosystem and its endemic species did an ecological rationale for protected area design take on more prominence within the U.S. Parks Service (Runte, 1979). Such a philosophy should also apply to the marine realm. Otherwise, new systems of marine protected areas may be designed on the basis of an out-dated approach to park design, one that secures areas for reasons other than the natural environment and then uses the natural environment as justification for its protection. Deep-sea corals have the potential to encourage the establishment of ecologically meaningful marine protected areas, and help shift marine conservation philosophy from a Grand Canyons to an Everglades approach.

This paper will discuss spaces-orientated deep-sea coral conservation. The objectives are (1) to identify and describe different techniques to protect deep-sea corals within marine reserves, (2) to demonstrate how deep-sea corals can function as a catalyst for the implementation of ecologically meaningful systems of marine protected areas, and (3) to use Nova Scotia's deep-sea corals as a case study for the described conservation approaches.

MARINE PROTECTED AREA STRATEGIES

Deep-sea corals are particularly well suited for protection within marine reserves, not only because of their ecological significance, vulnerability to disturbance, or importance as a conservation cornerstone, but also because they appear to be well protected by such a conservation strategy. Their benthic nature and strong affiliation with specific environmental variables of substrate and ocean depth permit them to be well-conserved within marine protected areas, even those reserves that are relatively small in size. Furthermore, since deep-sea coral communities can have distinct ecological boundaries, terrestrial approaches to the design of protected areas can often be applied to deep-sea coral conservation.

Several different strategies are available to identify those deep-sea coral communities in greatest need of protection. The planning approaches to be discussed in this paper include (1) representation, (2) ecological hotspots, (3) rarity, (4) productivity, and (5) species-orientated approaches (Table 1). These strategies were primarily developed for the terrestrial realm, but are directly applicable to deep-sea coral conservation. Such protected-area approaches, however, can be used not only to determine where a marine protected area would be best located, but also to justify why a given specific deep-sea coral location is in need of protection.

Regardless of which protected area approaches are utilized to conserve deep-sea coral biodiversity, the precautionary principle should be adopted. Such an approach would seek to protect deep-sea corals within reserves even before it is fully understood how best to achieve this goal, and recognizes that waiting for the unattainable goal of scientific certainty before a system of marine protected areas is implemented would be to knowingly permit the loss of coral habitats. Consistent with this approach is the concept of redundancy. By over-protecting deep-sea corals within a system of marine protected areas, coral areas will be buffered against gaps in scientific knowledge and from stochastic episodes that may degrade protected coral sites.

Representation

A representative approach to reserve design seeks to establish protected areas in all biogeographical units of a given area (Gauthier, 1992). In order for a site to be considered adequately represented, all types of enduring features typical of that biogeographical unit must be contained within established protected areas. Enduring features are components of a landscape that change over very broad time scales. On land, such features include floodplains, highlands and drumlin fields; in the ocean they include basins, ledges and submerged canyons. By protecting all types of representative enduring features within a given biogeographical unit, the chances of also protecting a full spectrum of species and habitats occurring within that unit are greatly enhanced.

Given the slow-growing, long-lived nature of deep-sea corals, and their importance as a foundation species for biological diversity, corals are arguably enduring features in themselves. If this is the case, then a representative approach to the design of a system of marine protected areas must target different types of coral assemblages. A given marine biogeographical unit that contains deep-sea corals could not be considered adequately represented until all types of habitats in which corals occur are contained within the boundaries of marine protected areas. Conserving one assemblage of deep-sea corals in one location would be unsatisfactory if other deep-sea coral assemblages with different dominant species or physical sub-environments were left unprotected elsewhere.

Table 1. Reserve design approaches applicable to deep-sea coral conservation

Protected area approach	Philosophy
Representivity	Protects enduring features typical of a given biogeographical unit
Ecological hotspots	Targets for protection the most ecologically significant areas
Rarity	Targets for protection areas containing rare species or rare spaces
Productivity	Conserves ecosystems with the highest productivity values
Single-species	Protects habitat of individual species

Ecological Hotspots

An ecological hotspot approach to reserve design targets those areas considered most ecologically significant. Such hotspots often have a disproportional ecological importance for their size. On land, approximately 44% of the planet's vascular plant species, and 35% of the vertebrate species, occur within 25 hotspots, totalling only 1.4% of the Earth's landmass (Myers et al., 2000). Similarly, coral reefs contain much of the planet's marine biodiversity despite occupying less than 1% of the world's oceans. By setting aside such important places, the total ecological benefit attained from a given system of protected areas is maximized by conserving the greatest number of species per area (Myers, 1990). Due to their heightened ecological significance, it is also rational to protect ecological hotspots even if an area has already been satisfactorily protected with representative reserves.

Although additional research is required to determine the importance of deep-sea corals as foundation species upon which other species rely, it is reasonable to suggest deep-sea coral areas are indeed ecological hotspots. They provide a structural complexity to the ocean floor with which several species appear to be correlated (Jensen and Frederiksen, 1992; Mortensen et al., 1995). By protecting deep-sea coral forests, therefore, a disproportionate ecological benefit for its size can be achieved for a given offshore system of marine protected areas.

Although all deep-sea coral assemblages are arguably ecological hotspots, some occurrences will undoubtedly be more important than others and should be targeted for protection first. Those coral forests that are (1) oldest, (2) largest, (3) most biologically diverse, or (4) least fragmented by anthropogenic disturbances, are the most important for protection. Coral assemblages that are "old-growth" will likely contain a greater structural and ecological complexity than younger occurrences. The largest assemblages are more likely to maintain a measure of stability and ecological integrity. Biologically diverse areas are important because they will contain the most coral and coral-dependent species, as well as the most spatially concentrated ecological interactions. Given the intensity of human use in the offshore, coral forests that have managed to survive in an unfragmented state are immensely important for conservation. They are ecological hotspots by default, simply because they remain intact. This does not, however, devalue their ecological significance in the marine environment; rather, it emphasizes the urgent need to protect those deep-sea coral areas that remain.

Rarity

The earth is currently experiencing an extinction spasm at a magnitude comparable to the loss of the dinosaurs at the end of the Cretaceous Period. Some estimates suggest 25 to 50% of all species present today will be extinct by the year 2100 (Soulé, 1991; Wilson, 1992). By targeting for protection, areas containing species vulnerable to extinction, the global rate of species decline can be slowed. A rarity approach to reserve selection would establish protected areas where rare species are known or expected to occur. With respect to deep-sea corals, this may include rare coral species or those rare species dependent upon corals as habitat.

Species are not the only entities vulnerable to extinction. Entire habitat types can become rare and go extinct. More than 99% of North America's tall-grass prairie has been converted to agricultural land (Groom and Schumaker, 1993). Limestone pavement ecosystems, known as alvars, are naturally rare spaces (Belcher and Keddy, 1992). Beringia, an arctic lowland tundra-steppe ecosystem, became extinct during the Quaternary Period as a result of climate change (Giterman et al., 1982; Hibbert, 1982). Given that deep-sea coral communities have been greatly damaged by anthropogenic disturbances, they are essentially rare spaces in themselves and can be targeted for protection on such a basis. Therefore, even if a given coral area contains no rare coral species or rare species dependent upon corals as habitat, it is still justifiable to protect deep-sea corals because such areas are spaces that are intrinsically rare.

Perhaps the best way to prevent rare species from becoming extinct is through prevention. By conserving areas through representative, hotspot and other approaches, species can be protected from extinction long before they become rare. This can only happen, however, if the proper areas are selected for protection and if the protected areas are of a sufficient size to maintain ecological integrity.

Productivity

Not all ecosystems are equally productive. A productivity approach to protected area design would locate the individual nature reserves of a system where productivity rates are highest. On land, such ecosystems include freshwater wetlands, salt marshes, and mangrove forests, while in the oceans, they include coral reefs, deep-sea vents, estuaries and upwelling zones. Productive ecosystems are important because they maintain a disproportional ecological significance for their size and provide ecological services important to the maintenance of the broader landscape beyond the spatial boundaries of the productive ecosystem itself.

More research is required to determine the productivity of deep-sea coral communities. Although the corals themselves accumulate mass very slowly, the true total productivity value of deep-sea coral communities is not only a measure of the total deep-sea coral biomass, but also of the biomass of species dependent upon the coral as habitat.

Regardless of their own intrinsic productivity values, deep-sea corals are sometimes located in productive areas of the marine environment. For example, they may occur in areas of strong upwelling where currents have exposed hard substrates that permit anchorage or where ample nutrients are provided to coral polyps. Such places are naturally productive, regardless of whether deep-sea corals are present. By targeting productive areas for the implementation of marine reserves, corals may be protected as byproducts.

Species-orientated Approaches

Protected areas can be located to help conserve specific species. In Canada, for example, Wood Buffalo National Park in Alberta, and Polar Bear and Woodland Caribou Provincial Parks in Ontario, target for protection the species that bare their names. A similar approach could be utilized in the marine environment for deep-sea corals. Single species approaches to conservation, however, have been criticized for ignoring landscape-level processes, for being too slow to address the biodiversity crisis, and for depleting a large proportion of resources allocated to biological conservation (Franklin, 1993). The importance of coral species for the maintenance of the broader marine environment, however, is undeniable. As such, deep-sea corals are an excellent target species for species-orientated conservation approaches in the offshore marine environment. They are keystone, umbrella, foundation and flagship species.

A *keystone species* is a species whose ecological role is of pivotal importance to a biological community and has far-reaching implications within the ecosystem in which it occurs. Keystone species include top predators, habitat modifiers, and symbiotic organisms. Deep-sea corals are keystone species in that they provide, on the ocean floor, a structural complexity upon which many other species depend.

An *umbrella species* is a species that, when protected, conserves other species as byproducts (Lambeck, 1997). For example, by setting aside from development an area sufficient to support a minimum viable population of a given far-ranging species, other species co-occurring within such a vast area will also be protected, regardless of whether or not they were the original focus of the systems planning. Deep-sea corals are umbrella species. By protecting them, other species will inevitably be protected as well.

Deep-sea corals are also *foundation species*. As such, they provide a physical structure on the ocean floor upon which other species depend. Targeting foundation species for protection is important because they provide habitat for many other species, thus helping to maximize the ecological benefits attained from a given system of marine protected areas.

Flagship species are those species that can be used to rally public support for conservation initiatives. They are usually visually appealing species that are politically correct for governments to protect. Species such as the tiger, panda bear, elephant, right whale, and manatee are a few examples. Deep-sea corals can, are arguably do, have public appeal. As such, they can be used as vectors to elicit support for the implementation of ecologically meaningful systems of marine protected areas.

Multi-use Approach

A multi-use approach to the design of a system of marine protected areas is desirable for deep-sea coral conservation. Since deep-sea corals cannot be entirely secured from degradation solely through the implementation of marine reserves, a broader approach that considers the entire marine environment is needed. Such an approach could zone areas surrounding no-take marine reserves to restrict anthropogenic activities that may degrade conserved coral areas but permit human activities that are less harmful. For example, although bottom-trawling should be excluded from buffer zones surrounding deep-sea coral marine protected areas, moderate levels of less damaging fishing techniques, such as longlining, might be permissible. In this regard, a *biosphere reserve model* is compatible with deep-sea coral conservation. A biosphere reserve is an area, recognized by the UNESCO Man and Biosphere (MAB) programme, where a balance is being sought between conservation and human use. Such a concept has more recently been expanded to include marine regions (Batisse, 1990; Kenchington and Agardy, 1990) and could readily be applied to deep-sea coral areas.

NOVA SCOTIA'S DEEP-SEA CORAL CASE STUDY

Nova Scotia's deep-sea corals occur primarily along the edge of the Scotian Shelf from the Northeast Channel at the Hague Line between Canada and the United States, to the Laurentian Channel at the provincial border with Newfoundland. These corals reside at depths between 100 and 600 fathoms, and are anchored to hard substrates of cobbles, boulders and bedrock outcrops (Breeze et al., 1997). Approximately a dozen species from the Orders Gorgonacea and Scleractinia have been identified thus far (Breeze et al., 1997).

Marine protected area strategies involving representivity, ecological hotspots, rarity, productivity, and focal species can all be used to protect Nova Scotia's deep-sea corals or to justify the protection of known coral occurrences. Most of Nova Scotia's corals occur within the (1) Scotian Slope, and (2) Bank Edges, Saddles and Channels marine biogeographical units, as identified and mapped by the Nova Scotia Museum of Natural History. The corals occur within three primary habitats: (1) channels, (2) submerged canyons, and (3) slopes (Breeze et al., 1997). Representative marine protected areas established on the Scotian Shelf, therefore, must contain deep-sea coral communities from all three of these sub-habitats in order to achieve satisfactory levels of representation. To do otherwise would be to exclude from protection significant enduring features of the Nova Scotian offshore.

Nova Scotia's coral hotspots are well known from the knowledge of local fishermen (Breeze et al., 1997). The localities with the oldest, largest, most biologically diverse, and least fragmented deep-sea corals are thought to occur within (1) the Northeast Channel between George's and Brown's Bank, (2) the Gully near Sable Island, and (3) the Laurentian Channel. These locations are often referred to as Hell Hole, Hell's Kitchen and Stone Fence, respectively. Other deep-sea coral hotspots may also occur on the Scotian Shelf, particularly within other submerged canyons, such as Dawson, Logan and Haldimand Canyons, as well as within other shelf channels. Further research and mapping is required to determine their true extent. Nonetheless, all three of these coral hotspots appear to be worthy of protection in their own right, regardless of representation.

Not enough scientific information currently exists for the Committee on the Status of Endangered Wildlife in Canada (COSEWIC) to rule on the status of Nova Scotia's deep-sea corals. Nonetheless, the coral forests themselves are essentially rare ecosystems and can be protected within marine reserves on such a basis. At such time as COSEWIC judges individual deep-sea coral species occurring within the waters off Nova Scotia to be vulnerable, threatened or endangered, localities containing such species should be given a heightened degree of importance for conservation.

In Nova Scotia, deep-sea corals are concentrated within channels, submerged canyons, and along the continental slope, where hard substrates have been exposed (Breeze et al., 1997). Such places are naturally productive because they contain strong currents and upwelling zones upon which many marine species rely. By protecting the productive areas of the Scotian Shelf, regardless of the presence of deep-sea corals, such species may be protected as byproducts simply because they occur where productivity rates are naturally high.

The responsibility for implementing a system of marine protected areas to conserve Nova Scotia's deep-sea corals lies primarily with Parks Canada and the Department of Fisheries and Oceans. A representative system of marine reserves falls under the jurisdiction of Parks Canada, while ecological hotspots, rarity, productivity and species-orientated approaches are primarily the responsibility of the Department of Fisheries and Oceans. No provincial legislation exists to protect corals in the offshore. Until such time as marine protected areas can be officially established under Canada's Parks Act or Canada's Oceans Act, interim protection in the form of fishing gear restrictions and temporary closures should be implemented under the Fisheries Act. Such an initiative would help safeguard deep-sea coral communities during the period when candidate sites have been identified for protection and such time as they can be formally secured through legislation.

CONCLUSIONS

Deep-sea corals are an important basal component of the offshore in need of protection within systems of marine protected areas. Representivity, hotspots, rarity, productivity, and focal species are all reserve design approaches that can be used concurrently to achieve this goal. Given the intensity, severity and frequency of the anthropogenic disturbances that threaten most deep-sea coral areas, the implementation of ecologically meaningful marine protected areas is of the utmost urgency. The longer we wait to establish such systems, the less likely deep-sea coral areas will remain intact for us to protect. To knowingly permit the decline and fragmentation of deep-sea coral areas would be to condemn forever the offshore to structural and ecological simplicity, and to diminish forever the rebound of species dependent upon coral as habitat. Marine protected areas can slow the loss of deep-sea corals and perhaps provide a path to a more sustainable use of the marine environment.

ACKNOWLEDGEMENTS

Funding for this research was provided by the Natural Sciences and Engineering Research Council of Canada, the Ontario Government, and the Biology Department of the University of Waterloo. Thanks to Dr. Martin Willison, Derek Fenton and the Ecology Action Centre.

REFERENCES

Batisse, M. 1990. Development and implementation of the biosphere reserve concept and its applicability to coastal regions, Environmental Conservation 17: 111-116.

Belcher, J.W. and P.A. Keddy. 1992. Protecting alvar vegetation: will the sum of the parts equal the whole? In J.H.M. Willison, S. Bondrup-Nielsen, C. Drysdale, T.B. Herman, N.W.P. Munro and T.L. Pollock (eds), Proceedings of the Science and Management of Protected Areas Association. Science and Management of Protected Areas Association, pp. 327-332.

Breeze H., D.S. Davis, M. Butler and K. Vladimir. 1997. Distribution and status of deep sea corals off Nova Scotia. Marine Issues Committee Special Publication Number 1, Ecology Action Centre, Halifax, N.S., 58 pp.

Engel, J. and R. Kvitek. 1998. Effects of otter trawling on a benthic community in Monterey Bay National Marine Sanctuary, Conservation Biology 12(6): 1204-1214.

Franklin, J.F. 1993. Preserving biodiversity: species, ecosystems or landscapes?, Ecological Applications 3: 202-205.

Gauthier, D. (ed.). 1992. Canadian Council on Ecological Areas framework for developing a nation-wide system of ecological areas: a strategy. Occasional Paper 12, CCEA.

Giterman, R.E., A.V. Sher and J.V. Matthews, Jr. 1982. Comparison of the development of steppe-tundra environments in west and east Beringia: pollen and macrofossil evidence from key sections. In D.M. Hopkins, J.V. Matthews, Jr., C.E. Schweger and S.B. Young (eds), Paleoecology of Beringia. Academic Press Inc., London, pp. 43-74.

Groom, M.J. and N. Schumaker. 1993. Evaluating landscape change: patterns of worldwide deforestation and local fragmentation. In P.M. Kareiva, J.G. Kingsolver and R.B. Huey (eds), Biotic interactions and global change, Sinauer Associates Inc., Sunderland, Massachusetts, pp. 24-44.

Hibbert, D. 1982. History of steppe-tundra concept. In D.M. Hopkins, J.V. Matthews, Jr., C.E. Schweger and S.B. Young (eds), Paleoecology of Beringia. Academic Press Inc., London, pp. 153-156.

Jensen, A. and R. Frederiksen. 1992. The fauna associated with the bank-forming deepwater coral *Lophelia pertusa* (Scleractinia) on the Faroe Shelf, Sarsia 77: 53-69.

Kenchington, R.A. and T. Agardy. 1990. Achieving marine conservation through biosphere reserve planning and management, Environmental Conservation 17: 33-44.

Lambeck, R.J. 1997. Focal species: multi-species umbrella for nature conservation, Conservation Biology 11(4): 849-856.

Mills, C.E. and J.T. Carlton. 1999. Rationale for a system of international reserves for the open ocean, Conservation Biology 12(1): 244-247.

Mortensen, P.B., M. Hovland, T. Brattegard and R. Farestveit. 1995. Deep water bioherms of the Scleractinian coral *Lophelia pertusa* (L.) at 64°N on the Norwegian Shelf: structure and associated megafauna, Sarsia 80: 145-158.

Myers, N. 1990. The biodiversity challenge: expanded hot-spots analysis, The Environmentalist 10(4): 243-256.

Myers, N., R.A. Mittermeier, C.G. Mittermeier, G.A.B. da Fonseca and J. Kent. 2000. Biodiversity hotspots for conservation priorities, Nature 403: 853-858.

Runte, A. 1979. National Parks: the American Experience, University of Nebraska Press, Lincoln, N.E.

Shafer, C.C. 1999. History of selection and system planning for US natural area parks and: beauty and biology, Biodiversity and Conservation 8(2): 189-204.

Soulé, M.E. 1991. Conservation: Tactics for a constant crisis, Science 253: 744-750.

Watling, L. and E.A. Norse. 1998. Disturbance of the seabed by mobile picking gear: a comparison to forest clearcutting, Conservation Biology 12(6): 1180-1197.

Wilson, E.O. 1992. The diversity of life. Harvard University Press. Cambridge, Massachusettes.

HOW TO PROTECT CORALS IN ATLANTIC CANADA

Mark Butler and Susan Gass
Ecology Action Centre, Halifax, Nova Scotia, Canada

CORALS IN ATLANTIC CANADA

About 20 species of gorgonians and stony corals have been identified from the waters off Atlantic Canada, gorgonians being the most conspicuous (Breeze et al., 1997). Two of the more commonly encountered gorgonians, which fishermen call trees, are *Paragorgia arborea* and *Primnoa resedaeformis*, suspension feeders found in waters with strong currents and an abundant supply of food. These corals attach to hard substrates, from rocky outcrops to isolated boulders, and appear to be confined to depths below 180 m. At such depths natural disturbance is likely to be limited; the main source of disturbance is human activities.

Human activities that do or could have an impact on corals in Atlantic Canadian waters are fishing, petroleum activities, cable laying, and sand and gravel extraction. In the past, shipwrecks and the dumping of munitions may have had some impact on deep-sea corals.

Climate change could have an impact on corals if it results in changes in ocean currents and temperatures, for example by slowing or even stopping the North Atlantic Oscillation (Hansen et al., 2001).

IMPACTS FROM FISHING

Mobile Fishing Gear

Fishing has the greatest impact on deep-sea corals in Atlantic Canadian waters because it is so widespread and so well established. Of all the gears, mobile or dragger gear has the greatest impact because it is heavy, is in constant contact with the ocean floor, and is towed for one to three hours. The target species are groundfish such as cod, haddock and redfish, and more recently shrimp. The gear consists of metal doors weighing up to 1000 kg, wire cables, plastic rollers or rubber rockhopper gear (rubber discs that bounce over the bottom), and the net. All of these components come in contact with the ocean floor; the width of the impact can extend from 50 to 100 m for gear commonly used in Atlantic Canada. When this gear meets emergent epifauna such as corals or sponges, there is no contest.

Reports from other countries of damage inflicted on coral by mobile gear (Fosså et al., 2000; Koslow and Gowlet-Holmes, 1998; Witherell and Coon, 2001), reports from fishermen and fisheries observers in Atlantic Canada, and comparisons of fishing activity and coral distribution in Atlantic Canada waters, make it clear that draggers regularly come into contact with corals (Breeze et al, 1997; DFO, 1998). The use of rockhopper gear has allowed draggers to fish in previously inaccessible areas such as canyons, a favoured habitat of corals.

(An American company markets a trawl door largely in use on the west coast of North America. The brand name is Canyonbusters; see www.net-sys.com/fish-buster.htm.)

Draggers in Atlantic Canada range in length from 15 to 50 m. Most of the vessels are in the 15 to 20 m range and target groundfish and shrimp. Representatives of this fleet have suggested that their vessels have little impact on corals because they cannot fish at the depths where corals grow. However, these boats can fish to depths of 250 to 300 m, well into the range at which *Paragorgia* and *Primnoa* are found. Draggers over 20 m can fish at even greater depths. Limiting factors for a dragger are the size of drum and amount of cable it can carry. There is a rough three to one ratio between length of cable and the depth to which a vessel can fish. For example, a vessel that carries 600 m of cable can fish to depths of 200 to 250 m.

It is difficult to determine the amount of coral lost to dragging. Breeze et al. (1997) recorded that fishermen note a loss of corals in several areas. Employment of ROVs or submersibles in areas thought to have corals may provide us with more information.

Dense concentrations of corals can impede fishing. When Breeze et al. (1997) interviewed Nova Scotian fishermen, they heard reports that draggers had in the past intentionally knocked down corals by towing a heavy chain between two vessels. Fishermen referred to this practice as "trimming the trees." Breeze and her colleagues did not encounter reports of recent intentional efforts to clear away corals.

Gillnets

This gear consists of rope, monofilament twine, weights and floats, and anchors. They are called gillnets because fish swim into the nets headfirst, and their gills get stuck in the mesh when they try to back out. Bottom gill-nets are anchored to the ocean floor at specific intervals, but still have a limited capacity to move back and forth across the bottom with the currents. The rope and mesh can entangle the corals and the anchors also inflict damage. Because it is not towed over the bottom, fixed gear has less impact than mobile gear on emergent epifauna. Small to mid-sized gillnetters, however, can set their gear in deeper water than draggers of a comparable size. Like all fishing gears, the level or intensity of fishing must be considered when evaluating impacts.

Groundfish gillnetters fish in areas where there are corals, and bring up corals in their nets. In addition, parts or sometimes entire sections of the net are lost and continue to fish or ghostfish as they roll over the bottom, potentially entangling corals and other marine life.

Longlines

Bottom longline gear is also fixed to the bottom by anchors. The gear consists of anchors, nylon line and baited hooks. The hooks can catch coral, the rope can wrap around corals, and the anchors can damage corals. Traditionally, longliners have fished close to, but not on dense aggregations of corals to avoid becoming caught in the corals and thus unable to haul in their gear. Relative to mobile gear, longline gear can be considered fairly benign, but in areas where there has already been extensive loss of corals from dragging, use of longline gear may need to be restricted. Assignment of observers to longline vessels, along with deployment of ROVs and submersibles, may help to determine the level of damage to corals from bottom longline gear.

Crab and Lobster Gear

With decline of the groundfish fishery in Atlantic Canada, new fisheries have developed for crustaceans in deeper waters. Lobster and crab gear consists of wire mesh traps attached to a groundline. The traps can damage corals and the line could wrap around the corals. As with other fixed gears, damage would likely be minimal, but would increase with the level of fishing activity.

Scallop Dragging

Scallop gear consists of a metal frame with a bag of attached metal rings that is dragged along the ocean floor. DFO (1998) reports that scallops are found from depths of 35 to 120 m on sand and gravel bottom. Corals (gorgonians and scleractinians) are usually found at greater depths and on bottom that contains more extensive hard substrate, such as boulders and rock faces. Therefore, scallop dragging is not considered a significant threat to corals, at least to the large, deep-dwelling gorgonians.

IMPACTS FROM OTHER OCEAN ACTIVITIES

Oil and Gas Drilling

Drilling for oil and gas can damage corals in two ways: by placing rigs and pipelines in areas where corals are present and through the discharge of drill cuttings (LGL Limited, 1998). Drill cuttings could smother corals growing at the base of a rig, while suspended sediment could interfere with the feeding and respiration of corals adjacent to the cuttings pile.

The discharge of produced water at the surface is unlikely to affect deep-sea corals. This would also apply to spills at the surface. In Atlantic Canada, as of May 2001, there has been no drilling at depths where large gorgonians are likely to be present. However, the entire edge of the Scotian Shelf from Georges Bank to the Laurentian Channel and over to the Grand Banks, a distance of 500 km or more, is now under exploration licence and it is likely that drilling at depths where corals are found will occur in 2002 or 2003.

Offshore waters in both Nova Scotia and Newfoundland are opened to oil and gas without any environmental screening or public consultation. Environmental assessments are conducted when specific projects, exploratory or production drilling, are proposed.

Cable Laying

Transatlantic cables could be laid across coral habitat. In depths of less than 1500 m the cable is buried to a depth of 1.5 m whenever possible. The total width of disturbance is described as 31 m and the impact as localized, limited and temporary (Jacques Whitford, 2000). Compared to activities such as dragging the impacts are likely not significant.

The laying of a transatlantic cable, however, is comparable to a transatlantic transect, and offers the opportunity to gather information on a range of subsea habitats. Unfortunately, cable companies do not appear to be taking advantage of these opportunities to collect biological data.

Seabed Aggregate Extraction and Mining

There is at present no seabed mining or aggregate extraction in Canadian waters, but government and the private sector have both expressed interest. Seabed mining could have an impact on deep-sea communities because some minerals of interest (e.g., magnesium nodules) may be found in deeper water. Aggregate extraction is unlikely to overlap with deep-sea corals because it is usually conducted in shallow coastal waters or on top of offshore banks.

CORAL PROTECTION

Because human activities are damaging the ocean floor, there is an immediate need to implement conservation measures to protect deep-sea corals and other sensitive benthic communities. In Atlantic Canada, several measures are possible. Marine protected areas (MPAs) are one option, but there are limitations to what MPAs can achieve. As a result, MPAs need to be accompanied by broader changes in ocean management. In the fisheries, for example, one such change would be gear substitution or modification, i.e., using fixed instead of mobile gear for groundfish. For damaging activities where there are no alternatives, the goal would be match the activity to less sensitive areas on the ocean floor or to exclude it completely.

Marine Protected Areas

Since corals are sessile animals, marine protected areas[1] (MPAs) may offer an effective means of protection. In Atlantic Canada, a series of MPAs would have to be established that would target areas of high coral abundance as well as critical source populations. MPAs, however, are heavily data-dependent and in Atlantic Canada we lack extensive knowledge of the distribution of corals and their relative abundance. MPAs are thus unlikely to be an effective way of protecting corals if we do not know where the corals are.

We do, however, have reports from fishermen of where they have encountered corals (Breeze et al., 1997). Based on these reports and other recent video

[1] Our definition of a marine protected area is simple: an area that is closed to some or all human activities.

footage obtained by the Department of Fisheries and Oceans (DFO), fishermen, scientists and conservationists acknowledge that there are at least two areas with a high concentration of corals, namely the Sable Island Gully and the Northeast Channel, particularly between Browns Bank and Georges Bank (Figure 1), that could both be declared MPAs. However, other measures need to accompany MPAs if areas with smaller concentrations of corals and the hitherto unidentified coral areas are to be protected.

If we proceed with establishing MPAs in areas of high coral concentrations, there are several options to consider. The first is a voluntary MPA (VMPA), which requires minimum government intervention. A successful VMPA requires heightened public awareness of the importance of deep-sea corals to marine ecosystems and biodiversity, and in particular of their critical role as fish habitat. Most importantly, this information needs to be communicated to the fishing, oil and gas, and mining industries so they can voluntarily avoid carrying out harmful activities in the protected areas. This can be done through notices to mariners, by publishing maps in local newspapers and relevant magazines, and by highlighting the areas in question on nautical charts, which are used by virtually all ocean users. VMPAs are an excellent way of promoting public awareness of the issue but the effectiveness of such a measure in protecting corals is difficult to determine, as VMPAs in no way provide legal protection.

A second possible measure for establishing protected or closed areas is through fisheries management legislation and specifically the federal *Fisheries Act*. The *Fisheries Act* has a provision that prohibits the damaging of fish habitat. Section 35 (1) of the act states that "no person shall carry on any work or undertaking that results in the harmful alteration, disruption or destruction of fish habitat." Fish habitat as described in the *Fisheries Act* refers specifically to "spawning grounds and nursery, rearing, food supply and migration areas on which fish depend directly or indirectly in order to carry out their life processes." Several studies have been carried out in different parts of the world that have shown deep-sea corals to be important habitat for fish, including commercially fished species (Fosså et al., 1998; Furevik et al., 2000; Witherell and Coon, in press; Heifetz, in press). In accordance with these studies and Section 35(1) of the *Fisheries Act*, coral areas could be closed to bottom-damaging fishing gear. An area closure under the *Fisheries Act* is a potentially rapid means of achieving protection for corals. If it has DFO support, fisheries managers could close an area overnight. However, closed areas under the *Fisheries Act* would not have the same permanence as those creased under MPA legislation and would only protect corals from fishing activities.

Canada's *Oceans Act* has specific provisions for the creation and implementation of MPAs. An MPA is defined in the *Oceans Act* as "an area of the sea that forms part of the internal waters of Canada, the territorial sea of Canada or the exclusive economic zone of Canada and that has been designated under the Act for special protection for one or more of the following reasons: (a) the conservation and protection of commercial and non-commercial fishery

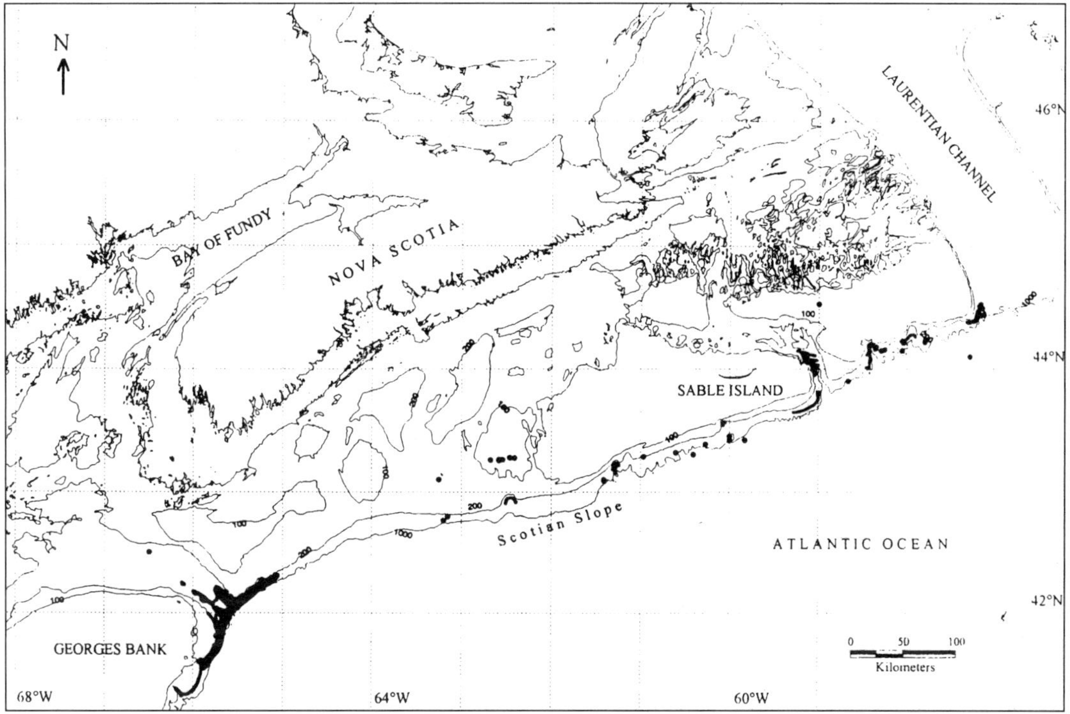

Figure 1. Distribution of corals (shaded areas) off Nova Scotia based on reports from fishermen, and museum and scientific collections of corals. (Reproduced from Breeze et al., 1997)

resources, including marine mammals, and their habitats; (b) the conservation and protection of endangered or threatened marine species, and their habitats; (c) the conservation and protection of unique habitat; (d) the conservation and protection of marine areas of high biodiversity or biological productivity; and (e) the conservation and protection of any other marine resource or habitat as is necessary to fulfill the mandate of the Minister."

An area with a high abundance of deep-sea coral could fall under one or more of these descriptions:

- Corals have been shown to act as important habitat for commercially fished species; by protecting corals as habitat the Minister is therefore fulfilling criterion (a).
- Corals have also been shown to be areas of high biological diversity (Fosså et al., 1999; Richer de Forges et al. 2000) and therefore satisfy criterion (d).
- Although a comprehensive assessment has not been carried out to determine the status of deep-sea corals off Atlantic Canada, reports from fishermen indicate that coral abundance has decreased in several areas (Breeze et al., 1997). Therefore, arguments could begin to be made to support criterion (b).
- As more research is carried out on deep-sea corals off Atlantic Canada, it may be revealed, for example, that *Lophelia pertusa* is only found in one area on the Scotian Shelf. Consequently, that area could be considered unique and protected under criterion (c). It could also be argued that corals themselves are unique organisms on the ocean floor as there are no other organisms of their size that provide the same extent of complex structural habitat. The term "unique" is not defined in the Act, therefore at this time the guidelines of what would fulfill criterion (c) are unclear.

The *Oceans Act* was passed in 1997, and as of 2001, no MPAs have been established under the *Oceans Act*. There are, however, several *Areas of Interest* in Atlantic Canada, two inshore and one (the Sable Island Gully) offshore. The Gully will become a fully protected MPA, but the government has not provided a formal date. It is unlikely that other offshore areas will become MPAs in the next few years. Therefore, this process has proven to be a lengthy one and most likely will not satisfy the immediate need to protect much of the deep-sea corals.

Although MPAs are one option for conservation, they should be considered but one important tool in ensuring the long-term existence of corals off Atlantic Canada. In addition to being heavily data dependent, MPAs generally protect only a small proportion of the ocean. New Zealand is one of the world's leaders in marine conservation and is currently trying to achieve protection of at least 10 percent of the coastal waters (Willison, 2000). This would leave the remaining 90 percent unprotected. Thus, although MPAs can be used to protect special areas of high coral abundance, they will leave the remaining ocean floor unprotected and vulnerable to impacts from resource exploitation. Furthermore, MPAs are difficult to enforce, especially when established in offshore areas, and unless preemptive measures are taken, the creation of MPAs can simply displace fishing effort to other areas, resulting in the rapid degradation of these other areas. MPAs are an important tool for marine conservation but they cannot protect the ocean on their own. There is a need for a comprehensive approach and a series of protection measures that will act cumulatively to protect marine species and habitats.

As more information is collected on the ocean floor we can then begin to work towards the ultimate MPA, which would involve zoning Canada's entire exclusive economic zone (EEZ). Based on our knowledge of the distribution of benthic organisms, the ocean could be divided into different zones depending on their vulnerability to human activities. We could then match human activities, such as fishing gears and oil and gas activities, with bottom types based on their degree of sensitivity (Watling and Norse, 1998). This integrated management option allows us to protect all sensitive benthic habitats including deep-sea coral communities. There are, however, several technical and organizational difficulties in mapping, classifying and zoning. These include the high cost of data collection; difficulties in choosing an effective classification system; the natural heterogeneity of the ocean floor, which makes it problematic to zone; and the lack of detailed knowledge on the impacts of human activities on ocean floor habitats — an absolute necessity if we are to match activities to appropriate zones. We would like to emphasize that an integrated management approach to ocean management can only be effective in protecting ocean floor habitats if conservation is given the highest priority throughout the process.

Further Measures to Protect Corals

To supplement MPAs, we would like to present another option for coral protection that we believe is also required. First, we would like to propose that fishing gears be ranked according to their impacts on the ocean floor and that the use of the least destructive gears be mandated for that fishery. For example, most groundfish can be caught using fixed gear, such as handline and longline gear, which are less destructive than mobile fishing gear, such as draggers (ICES, 2000). This practice will automatically reduce mobile fishing effort in Atlantic Canada and consequently reduce the surface area of the ocean floor that comes in contact with bottom-damaging fishing gear.

This option is particularly appealing now, while we still know little about the details of deep-sea coral distribution off Atlantic Canada — we could protect corals without having to know where they are. In addition, it would aid in the protection of all benthic organisms that provide structural habitat to the ocean floor and that are threatened by mobile fishing gear.

It has been our experience here in Atlantic Canada that increased public awareness is a big part of implementing successful conservation measures. The bottom of the ocean is often overlooked because it is so rarely directly seen. Deep-sea corals are charismatic species representing other benthic organisms inhabiting Atlantic Canada's ocean floor. In temperate waters, corals help bring the ocean bottom to life for the general public and elicit support for marine conservation and protection of the sea floor.

CONCLUSION

Deep-sea corals are threatened in Atlantic Canada by fishing (in particular dragging), as well as by oil and gas, and mining activities. MPAs offer one means of protection for corals and could be used to protect two well-known areas of high coral abundance, the Sable Island Gully and the Northeast Channel. At this time, it would be difficult to establish a system of MPAs to protect corals off Atlantic Canada, in part because we lack sufficient details about their distribution. This option therefore needs to be accompanied by broader changes in ocean management policies in order to protect the remaining unknown coral areas. In general, we are supportive of comprehensive measures such as integrated management or zoning of large areas of Canada's continental shelf and we advocate the restriction or phasing out of some of the more destructive ocean practices.

In closing, we would like to state that even in the presence of more comprehensive approaches to conservation (e.g., zoning), corals and the ocean at large will not be protected if we continue on our present path of industrializing the ocean and extracting renewable and non-renewable resources from it. Such activity will come at a cost – a highly degraded ocean.

LITERATURE CITED

Breeze, H., D.S. Davis, M. Butler and V. Kostylev. 1997. Distribution and Status of Deep Sea Corals of Nova Scotia. Marine Issues Committee Publication No. 1. Ecology Action Centre, Halifax, NS, 58 pp.

DFO. 1998. The possible environmental impacts of petroleum exploration activities on the Georges Bank ecosystem. Canadian Stock Assessment Secretariat Research Document 98/170. Department of Fisheries and Oceans, Ottawa, Ontario, 106 pp.

Fisheries Act, R.S.C. 1985, c. F-14, s.1.

Fosså, J.H., P.B. Mortensen and D.M. Furevik. 2000. *Lophelia* Coral Reefs in Norway: Distribution and Effects of Fishing. Institute of Marine Research, Bergen, Norway, 94 pp.

Furevik, D., L. Nottestad, J.H. Fosså, A. Husebo and S.B. Jorgensen. 2000. Distribution and concentration of fish in areas with deep sea corals. Abstract from the First International Symposium on Deep-Sea Corals. This volume.

Hansen, J., Turrell, W.R. and S. Osterhus. 2001. Decreasing overflow from the Nordic seas into the Atlantic Ocean through the Faroe Bank channel since 1950, Nature 411: 927-930.

Heifetz, J. In press. Coral in Alaska: Distribution, abundance, and species associations. Selected Papers Presented at the First International Symposium on Deep Sea Corals. Special Issue of *Hydrobiologia*.

ICES. 2000. Effects of Different Types of Fisheries on North Sea and Irish Sea Benthic Ecosystems: Review of the IMPACT II Report. Copenhagen, Denmark: International Council for the Exploration of the Sea.

Jacques Whitford Environment Limited. 2000. Report to Worldwide Fiber Inc. on Environmental Screening of the Hibernia Transatlantic Fiber Optic Cable. Jacques Whitford Environment Limited, Dartmouth, Nova Scotia, 70 pp.

Koslow, J.A. and K. Gowlet-Holmes. 1998. The Seamount Fauna off Southern Tasmania: Benthic Communities, their Conservation and Impacts of Trawling. Final Report to Environment Australia and the Fisheries Research Development Corporation, Australia, 104 pp.

LGL Limited. 1999. Environmental Assessment of Exploration Drilling off Nova Scotia. Canada-Nova Scotia Offshore Petroleum Board, Halifax, NS, 280 pp.

Oceans Act. 1996, Chapter C-31.

Richer de Forges, B., J.A. Koslow and G.C.B. Poore. 2000. Diversity and endemism of the benthic seamount fauna in the southwest Pacific, Nature 405: 944-946.

Watling, L. and E. A. Norse. 1998. Disturbance of the seabed by mobile fishing gear: a comparison to forest clearcutting, Conservation Biology 12: 1180-1197.

Willison, M. 2000. Marine protected areas: Canada's cold North Atlantic, Calypso Log, October 2000: 6-8.

Witherell, D. and C. Coon. 2001. Protecting Gorgonian corals off Alaska from fishing impacts. In Proceedings from the First International Symposium on Deep-Sea Corals. Ecology Action Centre, Halifax, NS.

THE ROLE OF THE CANADIAN OCEAN HABITAT PROTECTION SOCIETY IN DEEP-SEA CORAL EDUCATION AND CONSERVATION ADVOCACY IN NOVA SCOTIA

Derek P. Jones[1] and J.H. Martin Willison[1,2]
[1] Canadian Ocean Habitat Protection Society, Newellton, Nova Scotia
[2] School for Resource and Environmental Studies, Dalhousie University, Halifax, Nova Scotia, Canada

SUMMARY

The Canadian Ocean Habitat Protection Society (COHPS) was founded in 1996 by hook-and-line fishermen in southern Nova Scotia. The fishermen recognized the need for an organization dedicated to raising awareness of the importance of marine habitat for fisheries and to promoting public education about marine ecology. COHPS held many popular displays at fairs and gatherings, and continues with this work. The displays included "touch tanks" containing live fish and other local marine animals, and dried specimens of cold-water corals, which fishermen call "trees" and "bushes." The corals attracted a lot of attention because most visitors did not know corals existed near Nova Scotia. As a result, the fishermen of COHPS came in contact with teachers, professors, marine scientists and environmental activists, who were impressed by the coral specimens. This stimulated news stories, research projects, school visits and museum displays. As word spread, more specimens of deep-sea corals were gathered from fishermen and at wharves. Fishermen also provided local knowledge, which has been accumulated by COHPS and other cooperating groups. Internet communication facilitated the rapid growth of interest in the subject. The First International Deep Sea Coral Symposium was a product of the awareness initiated by COHPS. Using fishermen's knowledge, COHPS has identified the region of greatest coral diversity on the Scotian Shelf, and has proposed this as a marine protected area under Canada's Oceans Act. COHPS also continues to collect coral specimens from fishermen and to accumulate and interpret local knowledge, particularly as it relates to the distribution of deep-sea corals and their ecological role.

INTRODUCTION

In a landmark work published in 1997, Heather Breeze and co-workers reviewed existing knowledge of deep-sea corals in the Scotian Shelf region of eastern Canada, integrating fishermen's knowledge with information contained in formal scientific reports and museum records. The study by Breeze and others was supported by a grant to the Ecology Action Centre (EAC), a Halifax-based environmental advocacy group, from the Nova Scotia Museum of Natural History. The initial stimulus for this work came from contact between EAC and the Canadian Ocean Habitat Protection Society (COHPS), a fishermen's environmental advocacy group based in Cape Sable Island in southern Nova Scotia. Prior to contact with the fishermen of COHPS, the EAC was not aware of the existence of large gorgonian corals on Nova Scotia's continental shelf. Local fishermen have traditionally referred to these corals as "trees" and "bushes." The First International Symposium on Deep-Sea Corals was organized as a result of local recognition of the importance of the work by Breeze and co-workers (1997), which was itself founded in the work of COHPS and its constituent fishermen. Here we describe the history of the founding of COHPS, and its education and advocacy role in the Atlantic region of Canada.

FISHERIES AND HABITAT

During the 1990s, a series of fishery collapses in Atlantic Canada had severe socio-economic effects. Fisheries for cod and haddock, which had previously provided the economic foundation for many coastal communities, were either closed or curtailed. The socio-economic effects would have been even more serious if not for the fact that lobster, crab and shrimp fisheries gained in strength. In southern Nova Scotia, the lobster fishery replaced the cod fishery as the foundation of the regional economy. Although the total earnings of this restructured fishery are at least as great as before, employment levels are lower. The consequent economic and social inequity has led to societal malaise.

Down-turns in fisheries have been experienced for generations in coastal communities, but a collapse of the magnitude of that in the cod fishery had not been experienced previously on such a wide scale in Atlantic Canada. Fisheries in southern Nova Scotia were mostly affected less severely than in other parts of Atlantic Canada, but nevertheless the effect was felt throughout the region. By the mid-1990s it was clear to those working in fisheries that fish stocks were not recovering despite closures and reduced quotas, and that fundamental changes in marine community ecology may have occurred. This was fundamentally different from previous experience, in which economic down-turns had been caused mostly by market conditions.

Some of the fishermen who later founded COHPS were distressed but not surprised by the fishery collapses. For example, Sanford Atwood, a Clark's Harbour fisherman, had taken a piece of deep-sea coral to a local fishery officer in 1984 and had expressed concern that "the trees" were being seriously damaged by the use of mobile fishing gears (notably bottom trawlers using otter boards, see below). When he first formally approached the fishery officer about his concern, Sanford Atwood was making his living as his father had done — by using hooks attached to longlines to catch groundfish, particularly cod, haddock and halibut. Based on the observation that good catches of fish could be obtained where there were "trees," his father had taught him that "the trees" were important for groundfish. For this reason, Sanford's father taught him to fish among the trees, but to avoid damaging them. This requires skill and careful observation because fishing lines get caught in the corals; the trick is to try to lay the line near the corals, but not directly among them.

In view of the continuing damage being done to the coral trees, Sanford Atwood no longer fishes among the corals, having concluded that they are fishery incubators that should be protected. The story of his realization of this is intimately related to the story of the founding of COHPS.

There has been an uneasy relationship between hook-and-line fishermen and dragger fishermen in Nova Scotia for a long time. Ships using mobile fishing gears are known locally as "draggers" because the fishing gear (net or dredge) is dragged over the bottom. In the case of the otter trawl, the bottom-dwelling fish are caught by a very large and strong net that is attached to heavy steel components ("doors") designed to keep the net on the bottom and the mouth of the net open. Fishing gear of this kind is towed by powerful ships over considerable distances. As a result, a huge swath of altered habitat is left on the ocean floor each time the net is towed. Scallop dredges have a similar effect, albeit with a smaller footprint, and also give rise to controversy over impacts on benthic habitat.

When Sanford Atwood met the fishery officer in 1984, he explained that he regarded the trees as important habitat for his fishery. He expected the fishery officer to take this issue forward to his superiors as an important matter that needed urgent attention. Instead, he was ignored. Sanford Atwood was specially disturbed by this because there was a tradition of respect for the importance of "the trees" among hook-and-line fishermen, even though they had no formal scientific knowledge of deep-sea corals. By ignoring the trees, the fishery managers were ignoring more than one fisherman – Sanford Atwood's culture was being ignored.

THE FOUNDING OF COHPS

The Canadian Ocean Habitat Protection Society was founded by about a hundred Shelburne County, Nova Scotia, fishermen in 1996. The initial board of directors included Sanford Atwood, Derek Jones, Blantford Nickerson and Ronald Wolkins, whose combined fishing experience is varied and extensive, including work with handline, longline, mobile gear (bottom trawl) and trap fisheries, at all levels from captain to deckhand. The organization was created with the intention of meeting the following objectives: promoting public education about the importance of marine habitat in fisheries; advocating protection of marine environments against damaging human activities; and acting as a vehicle for environmentally aware fishermen to communicate with ocean managers, communications media, and other environmental groups. Some of the founding members had observed the purposeful destruction of deep-sea corals by use of chains and cables deployed from large vessels. This had been done to create "roads" along which nets could be dragged, or to "trim the trees" so that their nets would not be torn (Breeze et al., 1997). The COHPS members disapproved of such practices.

COHPS AND CORALS

Public Education

As a result of their experience that fishery managers were not interested in the importance to fisheries to deep-sea corals and other attached benthic marine invertebrates, the fishermen of COHPS decided to act independently of fishery managers by appealing to the public, and through the public to elected officials. In order to do this, they sought to improve their formal knowledge.

The most common gorgonian coral encountered by Nova Scotia fishermen is *Primnoa resedaeformis* ("red tree"). When the branches of this coral are cut across, the calcareous skeleton is revealed; it has alternating light and dark rings that look like the annual rings found in wood. Probably for this reason, a gorgonian coral specimen in the Archelaeus Smith Museum, on Cape Sable Island, was labelled as a "fossilized tree." The fishermen therefore approached a Dalhousie University marine geologist, David Scott, for more information. On seeing some specimens, Scott informed them that "the trees" were living corals, not fossilized trees. Derek Jones then turned to the Internet as a research tool, and quickly learnt a lot about corals, including the fact that tree-like deep-sea corals had first been reported hundreds of years ago.

The fishermen's discovery that "the trees" were actually living corals occurred at about the same time as COHPS was founded. The founders of COHPS felt that they had made an important discovery, and realized that ignorance of the true nature of "the trees" was widespread. They recognized that ignorance was a fundamental barrier to making progress towards sustainable fisheries, and that grass-roots education was a necessity. Their previous experience with government officials indicated that they would have to develop an independent public education strategy.

COHPS began its activities by organizing displays at local fairs, parades and other events at which local people gathered (Figure 1). In some cases, the display included a large "touch tank" containing living marine fish and other organisms that had been caught by the COHPS members and stocked

Figure 1. The Canadian Ocean Habitat Protection Society float at a parade in Shelburne, Nova Scotia

Figure 2. Specimens of Paragorgia arborea *and* Primnoa resedaeformis *mounted in a display case by Derek Jones to give the appearance of a coral forest on the sea floor. Small benthic invertebrates were placed among these corals to indicate the idea that the coral forest provides habitat for a community of organisms.*

in the tank. Children readily gathered around the tank, providing an audience for the members of COHPS to begin to explain the concept of marine habitat. The touch tank drew in an audience, providing the fishermen with an opportunity to deliver their message about the importance of protecting marine habitat. The touch tank created a stimulating yet relaxed educational environment.

COHPS also collected specimens of gorgonian coral trees and put these on display. Some of these have amusing local names, such as "bubblegum tree" for *Paragorgia arborea*. Derek Jones and Sanford Atwood made an attractive large portable display case in which specimens of these little known cold water corals were arranged so as to produce the appearance of a life-like "coral forest" (Figure 2). Large coral specimens were also put on display. The coral displays were used to explain that the ocean bottom is structurally complex because the living animals, such as corals, that occupy this space are themselves large and complex in shape. The fishermen found that many people were unaware of the existence of deep-sea corals and found them specially interesting. They also found that the concepts of marine habitat, and its susceptibility to damage, were readily conveyed when people took an interest in the corals on display.

Some of the coral displays were seen by local elementary-school teachers who invited COHPS to bring their displays to the teachers' schools (Figure 3).

Figure 3. COHPS members with school children and corals at a school in Shelburne County, Nova Scotia

Derek Jones improved the displays by adding additional specimens and by making posters. Schoolchildren sometimes took the coral conservation message home, and this sometimes resulted in additional contacts being made with fishermen, and new specimens being found. The recognition by teachers of the effectiveness of COHPS' educational approach was important in developing the self-confidence of the fishermen in their ability as educators.

Contact with the Ecology Action Centre

As a result of their community-focused education campaign, COHPS encountered Mark Butler of the Ecology Action Centre. Mark invited Derek Jones and Sanford Atwood to meet members of the Marine Issues Committee at the Ecology Action Centre. He also introduced them to Derek Davis, who had recently retired as Curator of Zoology at the Nova Scotia Museum of Natural History. As an expert in marine invertebrates, Derek Davis was familiar with the natural history of gorgonian corals, and with the museum records of their presence on the Scotian Shelf. Derek Davis encouraged the Ecology Action Centre to apply for one of the small research grants offered by the museum for the purpose of encouraging research relevant to the museum's focus on the natural history of Nova Scotia. The central element of the EAC's research proposal was the collection of fishermen's ecological knowledge of corals, and integration of this knowledge with museum and other scientific records. In this way, scientific collaboration was begun between EAC, COHPS and the Nova Scotia Museum of Natural History. The 1997 report by Heather Breeze and colleagues was the product of this research.

THE SECOND PHASE

Although COHPS and EAC are distinct organizations, in their work on corals they formed an effective rural-urban advocacy alliance. EAC is based in a city with a rich heritage of universities and research institutions and it was thus able to bridge the linguistic and cultural divide between the knowledge of fishermen and the knowledge of marine scientists. In so doing, environmental scientists and marine biologists such as Martin Willison (Halifax) and Don McAllister (Ottawa) were attracted to the issue, and it became possible to improve the quality and effectiveness of communication. Respect for the work of COHPS therefore continued to grow as a wider audience was engaged.

The Nova Scotia Museum of Natural History was particularly proud that its research grant had been used to such great effect, and the museum then created a permanent display of Nova Scotia corals in its main public gallery. Many of these specimens were originally collected by COHPS. An even larger display of COHPS' coral specimens was hosted by the Shelburne County Museum in the town of Shelburne during the summer of 1999. Shelburne is visited by many tourists, and the display attracted a steady stream of visitors, most of whom had no idea that corals existed in cold marine waters, such as those near Nova Scotia. Television, radio and newspaper reports were spawned as interest in the subject grew.

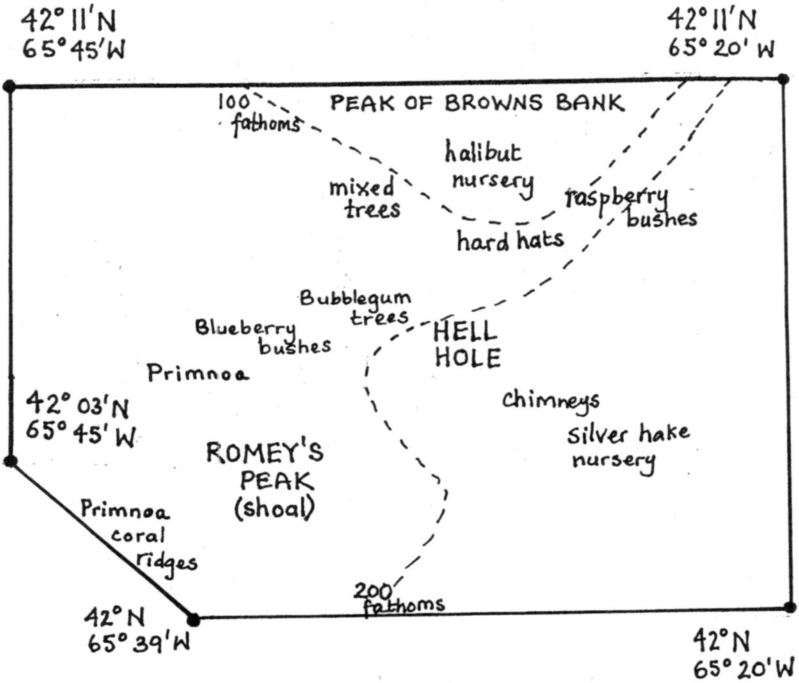

Figure 4. The COHPS proposal for a marine protected area in the Northeast Channel, south of Nova Scotia (see coordinates at corners). The locations of some of the valued components are shown in this diagram, which is based entirely on fishermen's knowledge (reprinted from Willison et al., 2001, with permission).

In order to improve communication, COHPS developed a website in collaboration with Coastwatch, a similar community-based organization in Maine (see: www.atlantisforce.org).

Marine Protected Area

One of COHPS' early proposals was for a marine protected area for deep-sea corals in a region known by local fishermen to have very dense stands of large gorgonian corals. Part of the area of concern is known locally as the "Hell Hole" because it is a very difficult place to fish, in part because of the presence of corals. The area lies where the Northeast Channel between Browns Bank and Georges Bank meets the slope at the edge of the continental margin.

Canada's Oceans Act provides the legal provisions for the establishment of marine protected areas to protect important fish habitat. It also permits marine protected areas to be established for marine areas with unique characteristics. The proposal for a coral forest marine protected area was originally made to the Department of Fisheries and Oceans in 1998, and was re-made in December 2000 by Martin Willison in a letter to Canada's Minister of Fisheries and Oceans. The COHPS proposal has been described in detail by Willison et al. (2001, and see Figure 4).

In 1999, the Canadian Hydrographic Service conducted a multi-beam sonar bathymetric survey of the region containing the proposed marine protected area. In 2000, shortly before the First International Symposium on Deep Sea Corals, the Department of Fisheries conducted a benthic survey in the area, and obtained videographs of gorgonian corals *in situ*. These were shown at the symposium. These government actions were a direct result of the foundation research and advocacy work conducted by COHPS.

Future Direction

Fisheries for groundfish have provided coastal communities in Nova Scotia with healthy livelihoods for generations. In the opinion of COHPS, these fisheries depend on the protection of marine habitat, including the habitat created by the large coral trees that have so fascinated generations of people in Nova Scotia. Following the symposium recorded in this proceedings, Sanford Atwood stated in a television interview that he would not give up fighting to protect the coral forests of Nova Scotia "as long as I live." He also said that it was time to "give something back to the ocean." COHPS continues to work in this spirit.

REFERENCES

Breeze, H., D. Davis, M. Butler and V. Kostylev. 1997. Distribution and status of deep sea corals off Nova Scotia. Marine Issues Committee Special Publication No. 1, Ecology Action Centre, Halifax, N.S.

Willison, J.H.M., D.P. Jones and S. Atwood. 2001. Deep sea corals and marine protected areas in Canada. In S. Bondrup-Nielsen et al. (eds), Protected

Areas in a Changing World, Science and Management of Protected Areas Association, Wolfville, N.S., Canada. In press.

NOVA SCOTIA'S CORALS AND CANYONS KIT: GETTING THE WORD OUT

Penny Doherty[1], Rayleen Brosha[1], Katherine Barrett[1], Melissa Watkins[1], Rosalie Galante[1], Derek Davis[2] and Mark Butler[1]
[1] Ecology Action Centre, Halifax, Nova Scotia, Canada
[2] Nova Scotia Museum of Natural History, Halifax, Nova Scotia, Canada

Researchers have identified ten species of horny corals (Gorgonacea) and ten species of stony corals (Scleractinia) in the waters off Nova Scotia. In recent years, fixed-gear fishermen have been raising concerns about the impact of mobile gear on the corals. The media and the public are becoming increasingly interested in corals. The Ecology Action Centre felt it was important to inform the public about the biology and ecology of corals, and the impact of human activities on these organisms.

Nova Scotia's Corals and Canyons is a multimedia kit produced for Grade 11 students in Nova Scotia. The kit contains general information on deep-sea corals and status reports on five coral species found off the coast of Nova Scotia. Video footage, slides, photographs and coral specimens are included. Some of the many species found in association with corals are profiled. Relevant articles on the threats to corals from fishing and petroleum activities are an important component of the kit. Several activities are outlined to enhance the understanding of important issues and to motivate students.

The kit is designed for a new high school course, Oceans 11, and was field-tested by two teachers of the course. The kit emphasizes the need for more research into all aspects of deep-sea corals. Recent research on deep-sea corals is recognized and material in the kit will be updated over time.

Since the 1870s fishermen have been pulling up various species of coral from deep waters off the coast of Nova Scotia. Coral was generally regarded as a nuisance and thrown back into the sea or taken home as a souvenir. It was frequently responsible for entangling nets and fishing lines, forcing fishermen to cut free their gear. More recently, fishermen have become concerned about the impacts of fishing gear on corals and subsequent effects on the groundfish fishery. Derek Jones and Sanford Atwood, Nova Scotian hook-and-line fishermen, were so concerned about the loss of corals that they started the Canadian Ocean Habitat Protection Society (COHPS). COHPS is a non-profit organization that educates people about the importance of marine habitat protection. Jones visits schools with his mobile coral museum to lecture on fishing gear impacts on marine habitat. He believes that the collapse of Canadian fisheries is directly related to the loss of coral. It was Jones and Atwood, along with other hook-and-line fishermen, who first brought their interest and concern about the state of corals to the attention of the Ecology Action Centre (EAC).

In 1997, the EAC was awarded a Rare Species Grant from the Nova Scotia Museum of Natural History to identify and map the distribution of deep-sea coral species on the Scotian Shelf and Georges Bank. The resulting report (Breeze et al., 1997) was based on information from fishermen, museum records, and the literature. Breeze mapped the distribution of five coral species from Georges Bank to the Laurentian Channel. Corals are distributed in canyons along the edge of the continental shelf and in deep channels between fishing banks. They are found on hard substrate from depths of approximately 180 m to 1000 m and deeper, in areas with strong current activity. The report identified two key areas for coral: "Hell's Kitchen" in the Sable Island Gully, and the Northeast Channel between Georges Bank and Browns Bank. Breeze et al.(1997) also suggested that ten species of horny corals (Gorgonacea) and ten species of stony corals (Scleractinia) may be found in the waters off Nova Scotia.

In 1998, three Bachelor of Science students, Melissa Watkins, Rosalie Galante and Katherine Barrett received funding from the Youth Environmental Challenge, a program of the Nova Scotia Youth Conservation Corps, to design and construct an educational kit about deep-sea corals off the coast of Nova Scotia. The kit, *Nova Scotia's Corals and Canyons*, combines information from government institutions, environmental organizations, researchers, museum collections, fishermen and scientific literature. The multimedia kit consists of a resource manual for teachers that presents general information on deep-sea corals, status reports on five deep-sea coral species found off the coast of Nova Scotia (Breeze, 1997), video footage, slides, photographs and various specimens of corals. Some of the many species found in association with corals and some of them are profiled in the kit. Relevant publications on the threats to corals from fishing and petroleum activities are an important component of the kit. Several activities are outlined to enhance the understanding of important issues and to motivate students.

In 1999-2000, EAC volunteers Rayleen Brosha and Penny Doherty did further work on the kit and had it field-tested by two teachers. Throughout the development of the kit, Derek Davis, a research associate at the Museum of Natural History, and Mark Butler of the EAC provided advice and reviewed the science in the kit. *Nova Scotia's Corals and Canyons* is available for loans of two weeks and is distributed by the School Loans Department at the Museum of Natural History in Halifax.

Nova Scotia's Corals and Canyons provides a rich assemblage of resources and learning materials and is designed to give students hands-on materials to work with. The kit will promote an awareness of corals and marine ecosystems and will encourage the involvement of students in long-term protection of the oceans. The kit is designed to give students a better understanding of marine environments by targeting youth in a high school curriculum that has not traditionally addressed marine environmental issues.

Development of the kit is part of an overall effort by the EAC to increase public awareness and knowledge of deep-sea corals. EAC members frequently speak to the public about corals and presented the kit at the Annual Conference of Nova Scotia Science Teachers in October 2000. Since 1997 the EAC has organized two local workshops to address ways of protecting corals on the Scotian Shelf. Participants included representatives of both the oil and gas and the fishing industries. A third workshop, with an emphasis on science, was held on 3 August 2000 following the First International Symposium on Deep-Sea Corals [see pp. 188]. The aim of the workshop was to develop a global strategy for protecting and mapping the distribution of corals.

Although originally designed as a resource for the Oceans 11 high school course, which had its first run as a pilot in September 1998, the kit has generated interest in elementary schools, universities and the general public. The response to *Nova Scotia's Corals and Canyons* has been positive. Like the marine environment, the kit has a dynamic nature. Recent developments in research into deep-sea corals are recognized and material in the kit will be updated over time to reflect current knowledge. The EAC continues to work with corals and to inform the public about the biology and ecology of these organisms and of the impact of human activities on them.

REFERENCE

Breeze, H., D.S. Davis, M. Butler and V. Kostylev. 1997. Distribution and status of deep-sea corals off Nova Scotia. Marine Issues Committee Special Publication Number 1. Ecology Action Centre, Halifax, N.S., 58 pp.

SPECULATIONS AND INTERSECTIONS: CANADA'S EASTERN CONTINENTAL SHELF

Susan Feindel
Independent Artist, Chester Basin, Nova Scotia, Canada

SUMMARY

In May-June 1999 I joined scientists from the Bedford Institute of Oceanography on a research cruise to the edge of the Scotian Shelf. I had sought this opportunity as part of my interest in the human relationship with land and sea. Coincidentally, I had listened to fishermen Derek Jones and Sanford Atwood speak on CBC radio about mysterious deep-sea corals, and thus I looked forward to observing the scientists in their work as they encountered these animals. Here I describe the development of my work in general, and more particularly the graphic images which arose from my experience on that cruise. These images were developed in conjunction with my Canadian National Art Gallery residency and subsequent studio production pursued as recipient of the National Art Gallery Claudia de Hueck Art and Science Research Fellowship. My work draws from mythology, fishing, military, medical and ecological concerns, and seeks to bring humans closer to issues connected with deep-sea coral habitats.

INTRODUCTION

I am a Canadian artist whose work is dedicated to developing images about the land we live upon. I am interested in land, past, present, and future, and its relationship to human habitation, and to the sea.

In 1995 I travelled with soil scientists and plant ecologists (the team of Dr. Larry Bliss, University of Washington, USA) to True Love Lowland on Devon Island, NWT, in the Canadian Arctic. I was inspired by outstanding evidence, in the form of permafrost patterns, of hidden cycles in our earth's surface. I felt, in the Arctic landscape, as though I were "inside my paintings." Settled again in Nova Scotia, I began at first to paint this motion of life in the earth, which I had studied in the Arctic. In *Ploughed Field* (Figure 1), I considered the ice-age patterns, and then imposed upon them the furrows of agriculture. This painting consists of iron oxides, carbon, earth pigments, and it links much of my work during the last decade.The transparency of the land's surface in this painting forwards the idea that when we are harvesting from the sea, the ecological effect can be similar to harvesting on agricultural land, or clear-cutting a forest.

In 1999, Nova Scotia inshore fishermen and activists Derek Jones and Sanford Atwood gave an impassioned plea on CBC radio for protection of the ocean habitat, the inshore fishery, and the little known, northern deep sea corals. Later, I met them giving a presentation at the Nova Scotia Museum of Natural History, surrounded by odoriferous skeletons of coral animals, col-

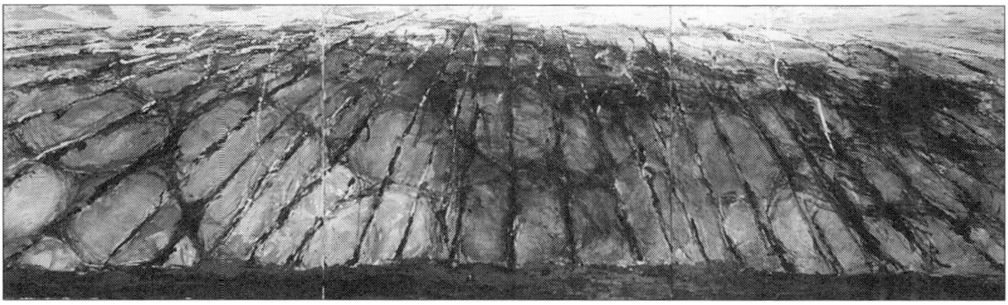

Figure 1. **Ploughed Field: Pattern**, *1998 (109 x 366 cm). Triptych, oil on canvas.*

lected from local fishing gear. I had recently been invited to accompany scientists of the Bedford Institute of Oceanography on their spring cruise aboard Canadian Coast Guard Ship *Hudson* (C.C.G.S. *Hudson*). Meanwhile, interest in northern corals had arisen in the scientific community. Federal funding became available to take the *Hudson* into areas on the eastern shelf edge known as the Gully, the Stone Fence, Hell's Kitchen, and Haldemand Canyon, in order to view these habitats by submersible video.

BACKGROUND

Generally, I want to reveal that extension of land which we live upon, but cannot see. The phrase "ocean clear-cutting" is sometimes applied to the benthic habitat by ecologists and inshore fishermen in relation to dragger/dredging activity. I am interested in this concept, and its relationship to my earlier work concerning the land. Unbridled agriculture, suggested by unending ploughed field images, were in many of the paintings which I showed in my exhibition *Figura*, mounted by the Ottawa Art Gallery in 1998. Of these, Petra Halkes, curator of *Figura*, writes in the book accompanying the exhibition:

> Through paint – with its propensities so similar to water, mud, blood, and excrement – Feindel seeks to recover the human kinship with the movement and historicity of nature, in order to encounter nature as a figure in her own right. (Halkes and Evernden, 1998, p. 12)

Northern Deep-sea Coral

Knowledge about these animals, known as "trees" by fishermen, comes with many claims: their height – 10 metres and more; their depth – 300-1000 m; their hardness – like stone; their form and colour – some a bloody, fluorescent red; their branches – like human organs; their lifespan - might predate the last ice age; their skeletal chemistry – close to human bone. Of special interest to me is the coral habitat. The places where the corals are found are the same Triassic-Jurassic outcrops that provide our fossil fuels. These are also the areas where my neighbours in Port Medway say "the big fish are."

Illustrations

Colour Plates 3 and 4 show two examples of several works painted from photographs taken of live corals retrieved from the deep. I wanted to depict the vulnerability of these animals, their presence in the brilliant light of day, and their displacement from their natural habitats. The illustration on the front of the book which depicts *Primnoa* in its natural habitat is another of these. The work, *Haldemand Canyon*, (Plate 3, top) was inspired by viewing this canyon on the research cruise.

Mining

Concurrent with fishing practices which disturb the ocean habitat near Nova Scotia, there has been an increase in oil and gas exploration. This has continued, and intensified since the oil crises of the mid 1970s – a forbidden fruit. The United States, which owns four-fifths of Georges Bank, extended its moratorium on drilling until 2012. But in 1999, when other countries such as Iceland and Norway were enjoying the benefits of choices that have protected their northern cod fisheries, Canada, on the Scotian Shelf, sold access to 2.2 million hectares of our offshore Canadian land to oil and gas corporations for exploration.

ARTIST'S BOOK

In the face of addressing these mammoth concerns, I created my third *artist's book*, a process which would frame my production on these topics in the future. An artist's book is an art genre in which the definition of a book, as we know it, is sometimes pushed to the limit. Often they are one-of-a-kind, or limited editions. My book, *Speculations and Intersections: Canada's Eastern Continental Shelf*, takes the position that knowledge is interdependent, and that in the case of the continental shelf, our knowledge can one day be regarded as mythology. My marine project has a personal focus, because my roots are in Nova Scotia, but personal too, is my choice to identify human bodily processes with ocean organisms, and my body with the sea. This science-informed art work, which positions scientific investigation in the arena of creativity, also places art in the arena of science and technology. It concurs with the writing of David Bloor in his book *Knowledge and Social Imagery* (1976).

David Bloor: Knowledge and Social Imagery

Bloor's radical claim in 1976 was that even the "hard sciences," such as physics and mathematics, are dependent on social factors. In his second edition, 1991, he responded to heated debates engendered by the first edition by presenting what he called his "strong programme," which he developed at the University of Edinburgh's Science Studies Unit.

Bloor states that a true process of scientific investigation must include both *internal history*, consisting of the collection and exploration of data, which is understood to be logical and repeatable, and *external history*, the substance of mythology, aesthetics, psychology and art, which many scientists believe cannot withstand methodologies applicable in the "hard sciences," such as mathematics and physics.

He states that a sociology of science can withstand the same rigour as the hard sciences. Conversely, scientific practice is strengthened by the rigorous inclusion of social factors. He asks the scientist to consider the connections that give rise to a system and status of thought. He requires that the methods used address all knowledge, equally.

Bloor's premise that the scientific method cannot be isolated from the sociological conditions which frame it, was evidenced aboard the research vessel. It is apparent to me that the ideologies of scientists are challenged in the dance between political, corporate and scientific interests. In the case of both art and science, where private corporate funding is available, the time, place, justification, and results of experimentation can be questioned.

Canadian handling of the northern cod stocks and, recently, several other fisheries, are examples that demonstrate Canadian political inability to manage our food resources. One of the major research departments at the Bedford Institute of Oceanography has seen a 60 percent reduction in funding and a 40 percent cut in scientific staff. This, at a time when the ocean habitat can no longer sustain Canadian fishing communities, and even less, the needs of people worldwide. These conditions place scientists in a position where the luxury of "arms-length" funding is more and more at risk. I am referring to that treasured policy of government or corporate expenditure, which protects intellectual freedom for artists and academics.

My visual material is yet another voice speaking on behalf of an environment that cannot speak for itself.

Description/Technique: *Speculations and Intersections: Canada's Eastern Continental Shelf*

There are twelve copies of the edition. It contains ten unbound pages and nine images. The text consists of about as many words per page. It is printed by an ancient intaglio technique in which ink is rubbed into the surface of a metal plate that has been etched with acid. The surface of the plate is then wiped clean, leaving ink in the fine lines of the design. A wet paper is placed on the surface of the plate, and both are passed through a printing press. Although understated visually, I believe that this subject warrants the care represented by this labour-intensive technique. Six of the figures are shown as Figures 2–7.

The thoughts and images that I compiled have passed through the filters of time, technology, politics and analysis. Each page represents a vast subject. Beginning (pp. 1-4) with images historical, sensual and anatomical that connect humans with the biology of the western Atlantic, the book then (pp.

5-7) evidences our connections technical, industrial and scientific, with the ocean habitat. Significant, is the convergence of activities where the Triassic/Jurassic formations are exposed at the continent's edge. There, too, is the habitat of northern deep-sea corals. My image of the corals is illusionary, offering a speculative glimpse of a habitat which the human eye will never see without simultaneously intruding on it. Adding to the filters of the imagination are those of the etched plate itself, a "leveller" of all references, which disallows weighted "truths" based primarily on advances in visual technology. Unvoiced is my speculation that apparent Canadian blunders in ocean management originate in corporate planning.

The Book itself:

The following is the text, page by page, of my artist's book, together with selected illustrations. I have revised page 10 (Technical Information and Acknowledgements) in a different format for the purpose of this paper.

Page 1. Illustration (not shown here). White on white embossed etching. Image and text copied from a book containing the woodcuts of Johann Zainer, 1499 (see acknowledgements).
Text:
Speculations and Intersections: Canada's Eastern Continental Shelf
Artist: Susan Feindel 1999
Saint Brendan (Circa 484-577) and His Monks at Sea,
Encounter a Siren, Meet a Holy Man Floating on the Sea,
And Their Ship is Attacked by Sea Monsters:
According to medieval legend, around the year 560, Saint Brendan and some Irish monks visited Astonishing countries in the Western Atlantic.

Page 2. Illustration (see Figure 2; detail): A filtering organism depicted within the grid of an agricultural furrow.
Text:
Suck in, Squeeze out, land and sea.

Page 3. Illustration (Figure 3; detail): Biological wonder and efficiency.
Text:
The Salp (Salpidae) propels and feeds itself in one motion.

Page 4. Illustration (not shown here). Sensual depiction of sea squirts in watercolour.
Text:
Urochordata: Ascidiacea and Salpidae:
Like Humans, the Sea Squirt and Salp have a notochord nervous system and gill stigmata.
Together, we are called Tunicates for the mantle which encloses us.

Page 5. Illustration (Figure 4; detail): Digital image recorded by acoustical sidescan aboard the C.C.G.S. *Hudson*.
Text:
At 75 meters, on Banquereau, sonar imaging, derived from Military surveillance, locates harvesting ruts 3 meters wide.

Figure 2. Illustration from page 2 of artist's book, detail

Page 6. (Reproduced in full as Figure 5.) To etch an image produced by wired technology, questions, in artistic terms, both the art and the science implied by the image. Can either effect change? The graphite line at the bottom of the page represents a horizontal drill path.
Sediment of the Grand Banks near Hibernia.
Text:
Drifting Floc: The stuff of diatom threads, plankton, mucus, copepods, and bentonite?

Page 7. (Reproduced in full as Figure 6.) Sediment of the Grand Banks near Hibernia.
Text:
Hibernia: N. 47°00', W. 48°35'
Sediment Stations 1-9 29/5/99

Page 8. (Reproduced in full as Figure 7.) A representation of a northern deep-sea coral habitat.
As described by Derek Jones, fisherman/activist, from Cape Sable, N.S.:
Coral gardens pulsate, blink, and glow, and are full of light. . . . At night

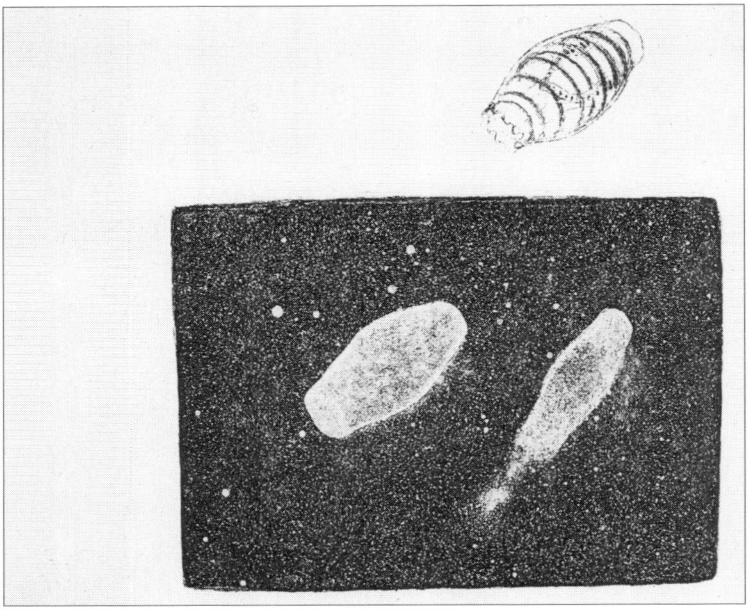

Figure 3. Illustration from page 3 of artist's book, detail

a deep sea coral brought to the surface sometimes will light up my boots.
Controversial descriptions of the deep, as seen by the light of bioluminescence, were recorded by William Beebe, in *Half Mile Down*, published in 1934.

Text: None

Page 9. No illustration.

Text:

Canaries of the deep, ancient coral animals, sustain a cradle habitat.

Page 10. No illustration.

Text:

Technical Information and Acknowledgements.

Epilogue

Colour Plate 4 (bottom) shows a reproduction of the work *When Activity Subsides, Life Blooms From Industrial Waste.*

Some of the forms in this painting are actual brass trailings from a casting pit. Their usage relates to a notion that in the world of imagination, when natural coral substrates are clouded in sediment, our metallic waste can be used

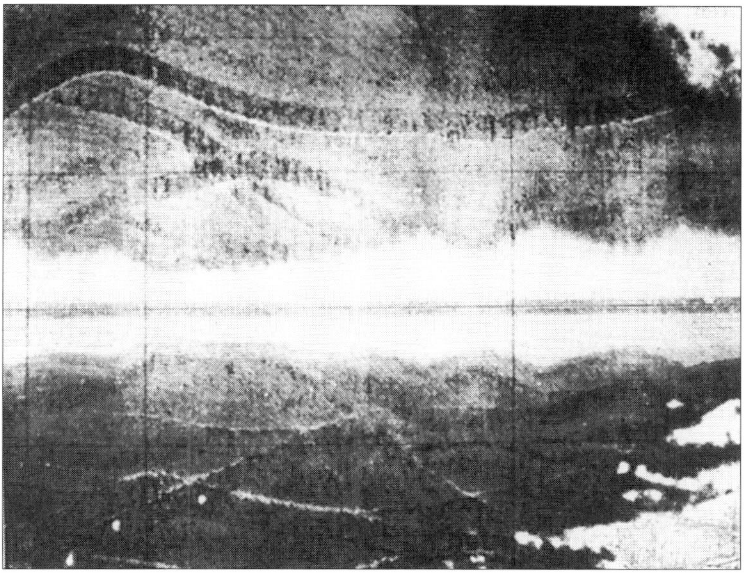

Figure 4. Illustration from page 5 of artist's book, detail

as a substitute. Projections based on the human practice of planting in rows, suggest that it will form the pattern of a tilled field in the sea, as on land.

ACKNOWLEDGEMENTS

Woodcut details and description, title page: Sankt Brandos, Ulm: Johann Zainer, 1499. 6.1 x 8.8 cm, 6.1 x 8.9 cm, 6 x 8 cm. Reproduced by permission of the British Library. British Library Shelfmark: In Saint Brendan 5.5741.

Hibernia floc image and Banquereau acoustical sidescan images: digitally recorded aboard research vessel C.C.G.S. *Hudson*, Bedford Institute of Oceanography, Dartmouth, N.S. Sediment collected aboard C.C.G.S. *Hudson*, May-June 1999, Grand Banks, Nfld.

Images of live coral Figures 2, 3, 5, are directly inspired by photographs taken by Mark Butler, Ecology Action Centre, Halifax, while aboard Sanford Atwood's boat.

Speculations and Intersections: Canada's Eastern Continental Shelf, was funded, in part, by the National Gallery of Canada 1999-2000 Claudia De Hueck Art and Science Research Fellowship.

Research was also made possible by the following: the creativity, patience and generosity of the scientists aboard the Bedford Institute of Oceanography vessel C.C.G.S. *Hudson*, May-June 1999 cruise; Dalhousie University pro-

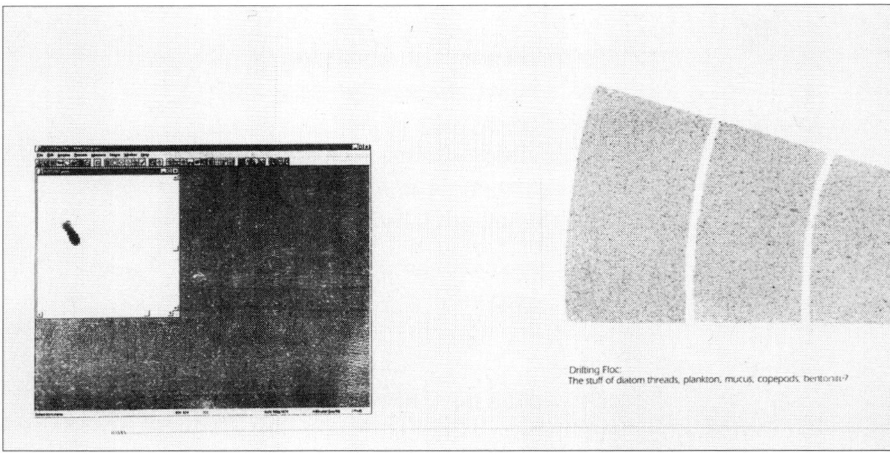

Figure 5. Page 6 of artist's book, reproduction in full

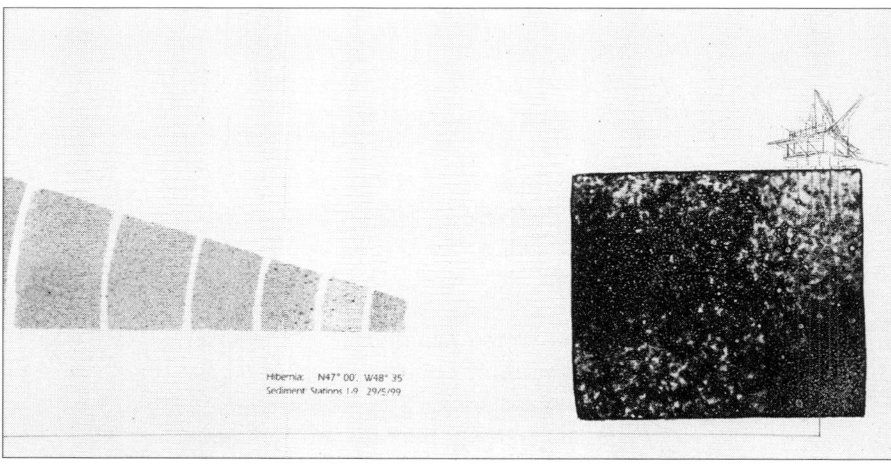

Figure 6. Page 7 of artist's book, reproduction in full

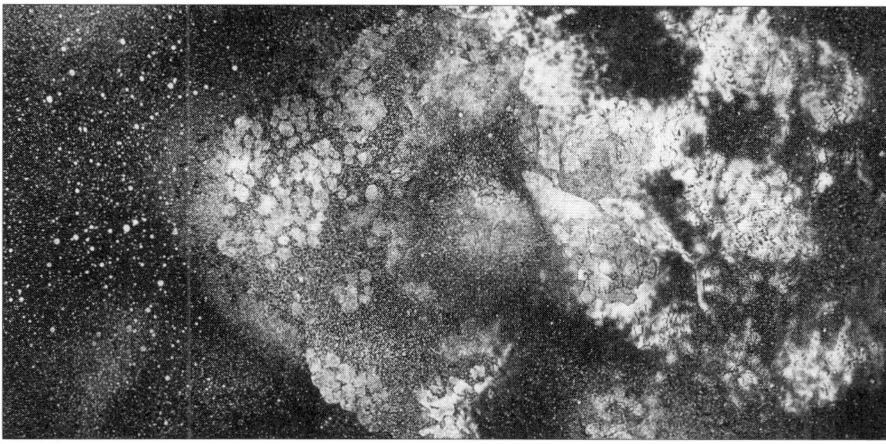

Figure 7.Page 8 of artist's book, reproduction in full

fessors, Dr. Martin Willison and Dr. Robert Fournier; Mark Butler, Marine Issues Coordinator, Ecology Action Centre, Halifax, N.S.; Dr. Derek Davis, Nova Scotia Museum of Natural History, Halifax, N.S.

For technical support, my appreciation to intaglio printmaker, Britt Quinlan of Paperwright, Ottawa, and to the Ottawa School of Art Print Room.

Finally, I wish to acknowledge the valuable insights of fishermen Derek Jones and Sanford Atwood, of Cape Sable, N.S., who are the stuff of legends.

REFERENCES

Beebe, W. 1934, 1951. Half Mile Down. Harcourt, Brace and Company, New York, 283 pp.

Bloor, D. 1976. Knowledge and Social Imagery. 2nd Edition. The University of Chicago Press, Chicago, 203 pp.

Halkes, P. and N. Evernden. 1998. Susan Feindel *Figura*. The Ottawa Art Gallery, Ottawa, 82 pp.

Zainer, J. Sankt Brandos. 1499. British Library Shelfmark. In Saint Brendan 5. 5741

DEEP-SEA CORAL CONSERVATION WORKSHOP

June Hall
Ecology Action Centre, Halifax, Nova Scotia, Canada

The goal of this one-day workshop, held on August 3, 2000, at King's College, Halifax, was to create the framework for a global conservation strategy for deep-sea corals. Moderator Martin Willison urged participants to focus on the big-picture issues, the coalitions to be made, not on local details.

The day began with two introductory presentations but thereafter followed a loose format designed to foster discussion. Participants spent much of the day in small groups, identifying concrete proposals for action. Topics covered were determined from tasks identified by Elliott Norse and Tony Koslow in their opening remarks. The day ended with presentations by representatives of each subgroup. The major achievement of the workshop was the creation of a new international organization dedicated to the conservation of deep-sea corals.

KEYNOTE PRESENTATIONS

9:00-9:30 am. Elliott Norse, President of the Marine Conservation Institute, Redmond WA. Strategic Planning for Deep-sea Coral Conservation: An American Perspective.

Dr. Norse began by stressing the need to bridge science and policy, to form liaisons with the general public and the decision-making community. He outlined two essential requirements:

1. *A systematic program of scientific research* to fill in the crucial gaps in our knowledge about cool-water/deep-sea corals, e.g., their distribution, ecology, the threats facing them, and the importance of their habitat for other species (including commercial species).

2. *A coordinated effort to promote conservation.* We also need to work out how to act for conservation at several levels and how to organize ourselves to do this effectively. We have the knowledge, the people, the funding and the institutions to help.

In the United States, he said, a number of laws affect corals:

i) *Magnuson-Stevens Act*: Corals affected if entering into commerce. Only a small number of genera are involved – precious corals used in jewelry.

ii) *Endangered Species Act:* Corals affected if under threat of extinction. Though there's a public process of nomination, decision-making and public comment, the ESA has been applied unevenly. It is effective for terrestrial and freshwater species, but weak for marine species; no marine invertebrate has ever been listed. ESA is not a good way to go.

iii) *Marine Sanctuaries Act:* Also imperfect – but has evolved. This act has regulated things such as oil and gas, but not fishing impacts. However, President Clinton's (26 May 2000) Executive Order to set up a national system of marine protected areas will offer more protection. There is to be an advisory group of scientists, fishermen, conservationists, etc.

This evolution of protection of corals from commerce —> endangered species —> own right, is a useful sequence. Protecting areas is the most robust strategy; you need far less specific knowledge than, for example, is required by CITES or the ESA.

We thus need to act at the level of threats, i.e., we need to restrict harmful activities. The *US Seabed Protection Act,* now before Congress, would protect certain areas against trawling and dredging. The bill won't pass, but is a useful exercise. Dr. Norse's chief advice: use all the contacts you have, and start at the top.

We also need a forum for communication. The proposed deep-sea coral list-serv will serve that purpose. Initial funding will come from the Marine Conservation Institute and from EAC.

Discussion

Several avenues of action were explored:

- Set up a new Web site (we have a volunteer but no money).
- Join the new ACES Web site (we could host one page).
- Join an existing list-serv with a broader mandate (we would be diluted).
- Set up own list-serv (will be acted on).
- Forward resolutions of the symposium, along with specifics and science, to CITES (inefficient, but Don McAllister will do).
- Make a submission to the MPA process (must be done before 25 August 2000. The group should discuss a concrete proposal with a dollar figure attached. Twenty million dollars have been designated for shallow-water corals. Elliott Norse will deliver the proposal to US senators.)
- Publicize deep-sea corals – to fishermen, the general public and scientists.

Tony Koslow, CSIRO Marine Research, Hobart. Strategic Planning for Deep-sea Coral Conservation: a Southern Hemisphere Perspective.

Southern Hemisphere seamounts are home to a diverse fauna including many endemic species. The major threat to these habitats is from the relatively new deep water fisheries, which now operate at all latitudes. Targeted species include orange roughy, found around seamounts at 600-1400 metres (trawling); alfonsino (mostly longline, some trawl); and Patagonian toothfish (mostly longline). The next bonanza is likely to be orange roughy living on seamounts in the Indian Ocean. Though we know little about this ocean, scientists realize we could lose much of its diversity.

Nevertheless, there is "movement at the station" at both the national and international levels. National initiatives include an Australian marine protected area (MPA) covering 15% of deeper, unfished seamounts south of Tasmania, and a New Zealand proposal to close 19 seamounts to fishing and to carry out a 5-year program of research.

Seamounts, whose fauna is often dominated by a variety of corals, pose an even bigger problem in international waters, where deepwater trawl fisheries are unregulated. Initiatives include one by Australia at the 2000 UNICPOA (UN Informal Consultative Process on Ocean Affairs) conference to establish high seas MPAs. The suggestion has received support from some EU nations, but not the US. A similar proposal by IUCN suggests an initial focus on MPAs for seamounts, vents, seeps and other sites with unique/threatened habitats. The process must include scientists, policy makers, conservationists and fisheries.

There's a crucial need for accurate maps, but the French have a huge amount of data on the South Pacific that should fill in many of the holes. There's also a need for regional cooperation in areas such as the west coast of North America.

Discussion

Topics included:

- the importance of sustainable use of habitats
- the long time-span that may be required to change fishing practices
- the many other organisms that live on and around seamounts, e.g., sponges
- whether high seas MPAs would work, even if international agreement was reached (e.g., ships registered under flags of convenience are unlikely to comply)
- precedents for international agreements, and three stages along the route to success: ignorance, denial and action.

SMALL-GROUP DISCUSSION

Work within four subgroups began around 11 am and continued after a break for lunch. A representative of each group then presented their findings.

International Cooperation Group

Alan Ruffman

This subgroup came up with a long list of suggestions on sources of funding, research topics, and avenues to be used to promote conservation of DSC.

1. *Europe:* Ideas for action should not be limited to the EU. Workshop participants were reminded that St. Pierre et Miquelon is part of France!

2. *Collaboration with existing projects:* e.g., Kim Conway's sponges off BC, Manfred Kreutter's program and also the Baja to Bering project (Grey Whales).

3. *Historical research:* In Canada we could draw from old charts, fishing reports, Russian surveys. Studies on the damage inflicted by dragging, such as Alex Rogers' work in Southampton, UK, would also be useful.

4. *Seeking funding for research:* A general principle: look for matching funds, or use cooperative programs to lever other grants. The Canadian government, for instance, will partner if you can get funds from elsewhere. Marathon Oil is inviting proposals to look at their data; European researchers have used these data already. Atlantic Canada Petroleum Institute another suggestion.

Today requests for research funding must often be couched in terms of their relevance to socio-economic issues (e.g., biomedical research, ecological significance, biological diversity).

5. *Potential collaborative programs:* We need to know more about the distribution of octocorals and *Lophelia* in the North Atlantic, and their relationship to oceanographic circulation. Cooperative projects, involving for example existing European programs, would help end the isolation of researchers.

Consideration should also be given to forming consortia (e.g., around sponges; Vancouver Aquarium with Parks Canada and GSC - Pacific).

6. *Identifying the gaps in our knowledge:* André Freiwald showed a map demonstrating where research had been done on North American corals, a useful tool for identifying gaps.

7. *Promotion:* Educate the public about threats to deep-sea corals. Make corals relevant to policy makers, funding sources.

Education/Media/Journals Subgroup

J. Murray Roberts

"We have plenty to help us get the story out: news, conflict and sexy corals! The question is, How to do it?"

1. *Through specialist scientific journals:* Thought to be less critical since review articles such as Rogers (1999) are already in the literature. It was felt, however, that the workshop should produce a declaration aimed at the scientific community, policy makers and stakeholders. A draft statement was read out and discussed, but so many questions were raised that those present agreed to postpone the process till after the meeting.

2. *Through other publications:* Paul Dayton discussed the article he was writing for TREE, and said he will try to get journals such as *Nature* and *Science* to comment on the symposium. He stressed the value of a cover illustration. *National Geographic* and like magazines should also be contacted and encouraged to do a story.

3. *Promotional activities:* The group emphasized the importance of publicizing corals in schools and other venues. NGOs such as WWF are especially useful. Scientists have lots of resources (photos, videos) that could be made available. Be active in promotion. Dabble in the waters of journalism! The new list-serv and the proposed page on the ACES Web site will be useful vehicles to get the word out.

Seamounts subgroup

Tony Koslow

This group concentrated on conservation strategies for seamounts located in high seas areas, though regional/national issues (e.g., progress in the Southern Hemisphere) were also mentioned. The group identified several routes of action:

1. *For the northeast Atlantic and the EU* - through the OSPAR Convention (Convention for the Protection of the Marine Environment of the North-East

Atlantic), which is reviewed annually. Much fishing is done in these areas on seamounts, e.g., increased take of orange roughy.

* Proposed action: Become involved in the next OSPAR round.

2. *For areas where there is no over-riding group* to orchestrate conservation efforts or to appeal to, e.g., the Indian Ocean.

* Proposed action: Try to put a consensus of scientific opinion in front of the UN General Assembly, where the oceans are discussed for one week each year.

3. *The International Seabed Authority* has so far involved itself only with mining potential. However, other topics such as cables are now on the ISA agenda.

* Proposed action: Attempt to have fishing and other harmful activities added to the ISA mandate.

4. *Through research:* Much remains to be learned about seamounts. For instance, do economically valuable species in the waters around seamounts also spawn and spend juvenile stages of their life cycle there – or are they just drawn to waters that are more productive because currents bring nutrients from elsewhere? What is the economic relationship between seamounts and fisheries?

North American Group

Elliott Norse

Three areas of action were defined:

1. *The Council on Environmental Cooperation (CEC)*, whose headquarters is in Montreal, recently declared protected areas as one of its priorities and set aside a fraction of its budget for the purpose. The Gulf of Maine/Georges Bank were identified as areas of concern.

* Proposed action: Propose a partnership with CEC and ask for research funding. Identify places where DSC occur, especially DSC communities in areas of concern.

2. *The U.S. Coral Reef Task Force*, established by President Clinton in June 1998 to lead the US response to the global crisis, should be contacted as soon as possible.

* Proposed action: Write a letter to the Task Force stressing that deep-sea corals and coral reefs are important and threatened, that there is a need for research funding, and that measures should be taken to protect DSC.

3. *Creation of a coalition*

* Action taken at the workshop: Founding of a coalition of interested people dedicated to the conservation of DSC. A long discussion ensued on the wording of the statement announcing the new group, the composition of the group, and its name. A show of hands determined the outcome in each case.

a) *Composition of the group.* The question was whether to restrict the group to scientists or to make it more broadly based. But as one speaker put it, Who could argue for a scientists-only group after *this* meeting? Finally, Elliott Norse proposed that the group use a language that speaks to diversity — that it be a broadly composed entity. Carried.

b) *Name of the group.* It took the meeting very little time to agree that the name of the new body should include the words "International Deep Sea Corals." Agreement on exactly what the new body should be — a group, coalition, society, association, forum or council — was again resolved by vote.

The winner: The International Forum on Deep Sea Corals (IFDSC).

c) *Final wording of the statement announcing the IFDSC:*

We, as scientists, managers, fishers, artists and other concerned citizens who have participated in the First International Symposium on Deep Sea Corals in Halifax, Nova Scotia, hereby join as the International Forum on Deep Sea Corals (IFDSC). The purpose of this coalition is to advance the science and conservation of deep-sea corals and their habitats through cooperation at the local, regional, national and international levels.

Martin Willison was named interim director of the IFDSC.

ABSTRACTS OF SYMPOSIUM PAPERS PUBLISHED IN *HYDROBIOLOGIA*

AGE, GROWTH AND RADIOMETRIC AGE VALIDATION OF A DEEP-SEA, HABITAT-FORMING GORGONIAN (*PRIMNOA RESEDAEFORMIS)* FROM THE GULF OF ALASKA

Allen H. Andrews[1], Erik Cordes[1], Melissa M. Mahoney[1], Kristen Munk[2], Kenneth H. Coale[1], Gregor M. Cailliet[1], and Jonathan Heifetz[3]

[1] Moss Landing Marine Laboratories, Moss Landing, CA, USA
[2] Alaska Department of Fish and Game, CWT and Otolith Lab, Juneau, AK, USA
[3] Auke Bay Laboratories, NMFS, NOAA, Juneau, AK, USA

Sustainable fisheries require 1) viable stock populations with appropriate harvest limits and 2) appropriate habitat for fish to survive, forage, seek refuge, grow and reproduce. Some deep-water habitats, such as those formed by deep-water stands of coral, may be vulnerable to fishing disturbance. The rate at which habitat can be restored is a critical aspect of fishery management. The purpose of this study was to characterize growth rates for a habitat-forming deep-sea coral. Two nearly complete colonies of red tree coral (*Primnoa resedaeformis*) collected from waters off southeast Alaska were used for an analysis of age and growth characteristics. CAT scans revealed that colonies consisted of multiple settlement events, where older basal structures provided for settlement of new colonies. The decay of total ^{210}Pb over the length of the colony was used to validate age estimates from growth ring counts. Age estimates were over 100 yr for sections near the heavily calcified base. Based on validated growth ring counts, growth of red tree coral ranged from 1.60 to 2.32 cm per year in height and was approximately 0.36 mm per year in diameter. These growth rates suggest that the fishery habitat created by red tree coral is extremely vulnerable to bottom fishing activities and may take over 100 years to recover.

THE DEEP-WATER CORAL *LOPHELIA PERTUSA* IN NORWEGIAN WATERS: DISTRIBUTION AND FISHERY IMPACTS

J.H. Fosså, P.B. Mortensen and D.M. Furevik
Institute of Marine Research, Nordnes, Bergen, Norway

The paper presents documentation on the distribution of, and damages to, deep-water reefs of the coral, *Lophelia pertusa*, in Norwegian waters. The reef areas have traditionally been rich fishing grounds for long-line and gill-net fisheries, and the coral habitat is known to support a high diversity of benthic species. Anecdotal reports claim that trawlers often use the gear, wires, chains and trawl doors to crush the corals and clear the area before fishing starts. To get an overview of the situation, information about the distribution and damage was collected from the literature, fishermen, and our own investigations. The results show that the corals are abundant particularly on the mid Norwegian continental shelf between 200 and 400 m depth. In general it seems that the largest densities are distributed along the continental break and at ridges of morainic origin. The reports from fishermen suggested severe damage to the corals and *in situ* observations using ROV confirmed the presence of mechanically damaged corals located on trawling grounds. A first estimate of the fishery impact indicates that between 30 and 50% of the reef areas are damaged or impacted. Fishermen claim that catches are significantly lowered in areas where the reefs are damaged. Potential ecological consequences of the destruction are discussed.

DNA SEQUENCES OF THE MITOCHONDRIAL *COI* GENE HAVE LOW LEVELS OF DIVERGENCE AMONG DEEP-SEA OCTOCORALS (CNIDARIA: ANTHOZOA)

Scott C. France* and Loretta L. Hoover
*Department of Biology, College of Charleston, Charleston, SC, USA

We are analyzing genetic diversity in deep-seamount octocorals with the ultimate goal of studying the effect of retention and dispersal of larvae on genetic population structure. Here we report on the sequence diversity of the mitochondrial *cytochrome oxidase* I (COI) gene among 11 species. Uncorrected pairwise sequence divergences ranged from 0.4% – 10.3% for comparisons among species spanning the intrageneric to interordinal levels. Relative to other invertebrates, these divergences are very low, suggesting that COI may not be useful as a genetic marker for studying dispersal among deep-sea octocoral populations. Possible explanations for the reduced rates of divergence observed include a lower rate of evolution for octocoral mitochondrial genomes and the presence of a gene, mt*MSH*, which may code for a mitochondrial DNA mismatch-repair system. We report the finding of mt*MSH* in three deep-sea octocorals (*Acanthogorgia* sp., *Corallium ducale*, and *Paramuricea* sp.), which brings the total published observations of this gene to six species, all in the subclass Octocorallia.

BEN FRANKLIN TEMPERATE REEF AND DEEP SEA 'AGASSIZ CORAL HILLS' IN THE BLAKE PLATEAU OFF NORTH CAROLINA

Robert Y. George
Department of Biological Sciences and Center for Marine Sciences, The University of North Carolina at Wilmington, Wilmington, NC, USA

This paper deals with two deep water coral ecosystems off the North Carolina coast, both under the depositional, erosional, and biological influence of the Gulf Stream and its eddies and episodic upwelling processes. The first coral community, Ben Franklin temperate reef, is located at 20 meters in Onslow Bay and is characterized by the ahermatypic coral *Oculina arbus-*

cula Verrill and unusually high abundance of the small predatory isopod *Eurydice bowmani* George and Longerbeam. The deep sea coral community, Agassiz Coral Hills, is located over the Blake Plateau at 650 meters, dominated by the fossilized, dead and living coral *Bathypsammia tintinnabulum*, with rare occurrence of the deep sea solitary coral *Flabellum goodei*. The deep sea coral sites were sampled with an otter trawl, outfitted with 'Benthos' flotation spheres. The results suggest that the fish and shrimp fauna exhibits high species richness in the Agassiz Coral Hills off North Carolina in comparison with a control site over the Blake Plateau off Florida. Species richness may be linked to episodic upwelling events along the Gulf Stream meanders. The fish fauna in the Agassiz Coral Hills includes the following two commercially important deep sea fish species: the wreck fish *Polyprion americanus* and the eel *Synaphobranchus koupi*. The Ben Franklin reef and the Agassiz Coral Hills are recommended as Marine Protected Areas.

CORAL IN ALASKA: DISTRIBUTION, ABUNDANCE, AND SPECIES ASSOCIATIONS

Jonathan Heifetz
Auke Bay Laboratory, Alaska Fisheries Science Center, National Marine Fisheries Service, NOAA, Juneau, AK, USA

To help identify fishery management actions that minimize the adverse impacts of fishing activities on corals in Alaska, the distribution and abundance of corals were analyzed based on trawl survey data collected during 1975 - 1998. We also examined the species of commercially managed fish that are associated with coral. Soft corals, primarily *Gersemia* sp. (=*Eunephthya* sp.), were the most frequently encountered corals in the Bering Sea. In the Aleutian Islands gorgonian corals, primarily in the genera *Callogorgia, Primnoa, Paragorgia, Thouarella,* and *Arthrogorgia* were the most common corals. In the Gulf of Alaska, gorgonian corals, primarily in the genera *Callogorgia* and *Primnoa*, and cup corals, primarily "Scleractinia unidentified," occurred most frequently. The Aleutian Islands area appears to have the highest abundance and diversity of corals. Some fish groups are associated with particular types of coral. Rockfish (*Sebastes* spp. and *Sebastolobus alascanus*) and Atka mackerel (*Pleurogrammus monopterygius*) were the most common fish captured with gorgonian, cup, and hydro corals, whereas flatfish and gadids were the most common fish captured with soft corals.

POTENTIAL CLIMATE SIGNALS FROM THE DEEP-SEA GORGONIAN CORAL *PRIMNOA RESEDAEFORMIS*

Heikoop, J.M.[1], Hickmott, D.D.[1], Risk, M.J.[2], Shearer, C.K.[3], Atudorei, V.[4]
[1] Los Alamos National Laboratory, Los Alamos, NM, USA
[2] School of Geography and Geology, McMaster University, Hamilton, ON, Canada
[3] Institute of Meteoritics, University of New Mexico, Albuquerque, NM, USA
[4] Department of Earth and Planetary Sciences, University of New Mexico, Albuquerque, NM, USA

The deep-sea gorgonian coral *Primnoa resedaeformis* has an arborescent skeleton composed of both calcite and a horn-like structural protein called gorgonin. We have investigated potential climate records in corals from Alaska, the eastern seaboard of Canada and the United States, and a Southern Ocean (Pacific sector) seamount. Temperatures at these sites range from 4 – ~10°C.

$\delta^{18}O$ values of the calcite show strong evidence for isotopic disequilibrium. Extraction of $\delta^{18}O$ paleotemperatures is therefore not straightforward. Sr/Ca data, analyzed by Secondary Ion Mass Spectrometry (SIMS), suggest that temperature might be a control on calcite Sr/Ca in *Primnoa resedaeformis*, but that growth-related kinetic effects could also be important.

Based on previous $\delta^{14}C$, $\delta^{13}C$ and $\delta^{15}N$ measurements, it has been suggested that particulate organic matter (POM) from the surface is an important carbon source to the polyps and the gorgonin fraction of the *Primnoa* skeleton. $\delta^{15}N$ and $\delta^{13}C$ of polyps and gorgonin show similar regional differences to $\delta^{15}N$ and $\delta^{13}C$ of surface POM. Polyps and contemporaneous gorgonin correlate strongly for both $\delta^{13}C$ and $\delta^{15}N$. The influence of nutrient isotopic composition and climate and productivity variations on the isotopic composition of surface POM may therefore be recorded in gorgonin layers.

These corals have very long lifespans (several centuries). The potential exists, therefore, to obtain extended records of surface productivity, deep ocean temperature and chemistry of value to climatologists and fisheries managers.

MAPPING AND IMAGING DEEP-SEA CORAL REEFS OFF NORWAY, 1982 - 2000

Martin Hovland, Steinar Vasshus, Arne Indreeide, Leslie Austdal and Øivind Nilsen
Statoil, Stavanger, Norway

The survey and mapping group (SMG) of Statoil is responsible for all seafloor mapping for pipelines and field development in Statoil. During numerous reconnaissance and pipeline route surveys over large portions of our contintental shelf, in the North, Norwegian, and Barents Seas, we have only detected deep-water coral reefs in a few specific areas. The first reef we found was in 1982, using a combined side scan sonar and sub-bottom profiler system off Fugløy, northern-Norway. Thereafter, numerous reefs were mapped with similar systems off mid-Norway, during the period 1985 - 1990 for the reconnaissance and final route mapping of the Haltenpipe project. During 1997 we also mapped some previously known reefs in the Trondheimsfjord area while surveying a route for the Tjeldbergodden - Skogn pipeline project. Between 1997 and 2000 we have mapped more reefs along pipeline routes on the outer mid-Norway continental shelf. These reefs are of a smaller size (less than 5 m high) than those mentioned above (5 to 31 m high).

Although side scan sonar and sub-bottom profilers provide a less ambiguous detection of reefs, we have found that modern multi-beam echosounder data can provide adequate remotely sensed data for deep-water coral reef mapping. The interpretation of the reefs is based on visual documentation by ROV (remotely operated vehicle) or by sampling with gravity corers and grabs (ground-truthing). Based on a limited amount of such ground-truthing, it has been possible to extrapolate and use the specific morphological characteristics of the reefs to map their density and distribution.

For the Haltenpipe project we mapped an offshore route corridor of about 200 km length at a width of 3 km with multi-beam echosounder. On the basis of ground-truthing 14 of the suspected coral reefs, we have found the total number of reefs to be 57 within the 600 km^2 mapped area. All these reefs are higher than 5 m (the highest is 31 m) and of diameters at their base of more than 50 m. Although they occur in local clusters with up to 10 reefs per km^2, the mean density of reefs along the entire (200 km long) transect is only 0.09 suspected reefs per km^2. However, there is a large regional density variation, with the highest regional density being 1.2 reefs per km^2 in an area of sub-cropping Palaeocene sedimentary rocks.

A brief discussion of why the corals have constructed their reefs in the deep, cool, and generally "hostile" waters of the Norwegian continental shelf and fjords concludes with them probably subsisting on a reliable and steady nutrient source, independent of season and variations in the Atlantic Drift ("Gulf Stream"). This positive environmental component is called "hydraulically active substrata." It is thus speculated that micro-organisms, bacteria etc., utilizing the hydraulically activated chemical porewater gradients, cause a local enrichment on which the cnidarian organisms ultimately depend.

DISTRIBUTION AND ABUNDANCE OF FISH IN DEEP-SEA CORAL HABITATS

Å. Husebø[1], L. Nøttestad[1]*, J.H. Fosså[1], D.M. Furevik[1] and S.B. Jørgensen[2]

[1] Institute of Marine Research, Nordnes, Bergen, Norway

[2] University of Copenhagen, København, Denmark

Experimental fishing with long-lines and gillnets was conducted on the continental shelf off Southwestern Norway between 150 and 300 m depth. Abundance and distribution of redfish (*Sebastes marinus* L., 1758), ling (*Molva molva* L., 1758), and tusk (*Brosme brosme* Ascanius, 1772) were quantified in *Lophelia pertusa* (L., 1758) coral reefs and in non-coral habitats. The largest catches of redfish were made with long-line fleets set in coral reef habitats. Ling and tusk were also most numerous in coral habitats, although not statistically significant. Fish caught in coral habitats tended to be larger in size than in non-coral habitats. The diet of redfish, tusk and ling included the same prey groups in all habitats, but they differed at the species level. *Lophelia*-reefs may provide a profitable feeding place for tusk. For the planktivorous *Sebastes*, on the other hand, their affinity to the reefs seems primarily to be related to the physical structure offered by the reefs.

MEGAFAUNA ASSOCIATIONS WITH DEEPWATER CORALS (*PRIMNOA* SPP.) IN THE GULF OF ALASKA

Kenneth J. Krieger and Bruce L. Wing
Auke Bay Laboratory, Alaska Fisheries Science Center, National Marine Fisheries Service, NOAA, Juneau, AK, USA

Few *in situ* observations have been made of deepwater corals and therefore little is known about their biology or ecological significance. Deepwater corals (*Primnoa* spp.) were observed from a manned submersible at 11 sites in the Gulf of Alaska from 1989 to 1997 at depths of 161-365 m. We identified 10 megafaunal groups that associate with *Primnoa* to feed on the coral, use the coral branches for suspension feeding, or for protection. Predators on *Primnoa* polyps included sea stars, nudibranchs, and snails. Sea stars were the main predators, consuming 45% and 34% of the polyps at two sites. Suspension-feeders included crinoids, basket stars, anemones, and sponges. Most suspension-feeders observed at depths >300 m were associated with *Primnoa*. Protection seekers included rockfish, crab, and shrimp. Six rockfish species were either beneath, among, or above *Primnoa*. Shrimp were among the polyps, and a pair of mating king crabs were beneath *Primnoa*. These observations indicate *Primnoa* are important components of the deepwater ecosystem and removal of these slow-growing corals could cause long-term changes in associated megafauna.

THE FEASIBILITY OF TRANSPLANTING BLACK CORAL (ORDER ANTIPATHARIA)

Anthony D. Montgomery
The Oceanic Institute, Waimanalo, Hawaii, USA

Is transplanting black coral feasible? In order to test the feasibility of transplanting black coral experimentally, forty black coral fragments were transplanted in five locations across Hawaii. One fragment, *Antipathes ulex* (Ellis & Solander), was transplanted at Makua, Oahu, Hawaii, next to the parent colony. Ten fragments of *Antipathes dichotoma* (Pallas) were transplanted at Kahuku, Hawaii, Hawaii, next to the parent colony and nine fragments of *A. dichotoma* (Pallas) were transplanted from Kahuku, Hawaii to Honokohau, Hawaii, Hawaii. Ten fragments of *A. dichotoma* (Pallas) were transplanted at two different locations each at Makena, Maui, Hawaii, ten next to the parent colony and ten away from the parent colony. The mean height of the fragments was 11.8 cm (SD ±3.7) at time of transplantation. The mean height of all surviving fragments was 5.1 cm (SD ±3.1) approximately 24 months after the initial survey. The mean survival was 45%. The mortality observed in this study was in the form of missing fragments. All observed fragments were alive. Based on the results of this study, the transplanting process needs to be refined. The microhabitat chosen for transplantation is critical. Further experiments are recommended to test for a minimum size required to transplant while minimizing the impact on the donor colonies. Also, the advantages and disadvantages of transplanting as well as the important criteria in determining the need for transplanting are discussed. A flow chart is given to show the decisions that need to be made before transplanting black coral.

DEEP-WATER *OCULINA* CORAL REEFS OF FLORIDA: BIOLOGY, IMPACTS, AND MANAGEMENT

John K. Reed
Harbor Branch Oceanographic Institution, Division of Biomedical Marine Research, Fort Pierce, FL, USA

Deep-water *Oculina* coral reefs, which are similar in structure and development to deep-water *Lophelia* reefs, stretch over 167 km (90 nmi) at depths of 70-100 m along the eastern Florida shelf of the United States. These consist of numerous pinnacles and ridges, 3-35 m in height. Coral growth rates average 16.1 mm yr^{-1} and biodiversity is very rich. Extensive areas of *Oculina* rubble may be due to human impacts (e.g., fish trawling and dredging, anchoring, bottom longlines) and natural processes such as bioerosion and episodic die-off. Early in the 1970s, the reefs were teeming with fish. By the early 1990s, both commercial and recreational fisheries, including scallop, shrimp, grouper, snapper, and amberjack, had taken a toll on the reefs and especially on populations of grouper and snapper. A 315 km^2 (92 nmi^2) area was designated the *Oculina* Habitat of Particular Concern (HAPC) in 1984, prohibiting trawling, dredging, bottom longlines and anchoring, and legislation was enacted in 2000 for expansion of the *Oculina* HAPC to 1029 km^2 (300 nmi^2). The United States Coast Guard has been charged with surveillance and enforcement of the ban on bottom fishing and trawling. The primary difficulties in protecting these reefs and other deep-water Marine Protected Areas are their remoteness and time required to engage an enforcement vessel. Education regarding the nature and importance of these rich resources is important for better self regulation and surveillance by the fishing community. Only by bringing deep-water reefs to the public, the fishing community, and enforcement agencies, through video, photos, and education will there be better understanding and acceptance for the need of protection for these unseen resources. This paper reviews the current knowledge on the deep-water *Oculina* reefs, including the biology, geology, human impacts, and history of conservation and management.

COMPARISON OF DEEP-WATER CORAL REEFS AND LITHOHERMS OFF SOUTHEASTERN U.S.A.

John K. Reed
Harbor Branch Oceanographic Institution, Division of Biomedical Marine Research, Fort Pierce, FL, USA

Two types of deep-water coral bioherms occur off the coast of southeastern United States: *Oculina* and *Lophelia/Enallopsammia*. The deep-water *Oculina* bioherms form an extensive reef system at depths of 70-100 m along the shelf edge off central eastern Florida. These reefs are comprised of numerous pinnacles and ridges, 3 to 35 m in height. Each pinnacle is a bank of unconsolidated sediment and coral debris that is capped on the slopes and crest with living and dead colonies of *Oculina varicosa*, the ivory tree coral. In comparison, deep-water reefs of *Lophelia pertusa* and *Enallopsammia profunda* corals occur at depths of 500-850 m (maximum 150-m relief) along the base of the Florida-Hatteras slope in the Straits of Florida. On the western edge of the Blake Plateau off South Carolina and Georgia, 54-m high banks of *Enallopsammia* and *Lophelia* occur at depths of 490-550 m, whereas on the eastern edge of the plateau the reefs form structures 146 m in height and at depths of 640-869 m. The geomorphology and functional structure of both the *Oculina* and *Lophelia* reefs are similar. North of Little Bahama Bank, at depths of 1000-1300 m, a region of bioherms is dominated by the coral *Solenosmilia* sp.; *Lophelia* is reportedly absent. This paper summarizes 20 years of submersible studies on the deep-water *Oculina* reefs, describes submersible reconnaissance of deep-water *Lophelia* reefs off the southeastern United States, and contrasts these types of bioherms with the deep-water lithoherms in the Straits of Florida west of the Bahamas.

INFLUENCE OF TEMPERATURE AND NUTRIENTS ON SPECIES RICHNESS OF DEEP WATER CORALS FROM THE WESTERN COAST OF THE AMERICAS

Héctor Reyes-Bonilla[1] and Gabriela Cruz-Piñón[2]
Universidad Autónoma de Baja California Sur, Departamento de Biología Marina, La Paz, B.C.S., México.
New addresses:
[1] Division of Marine Biology and Fisheries, Rosenstiel School of Marine and Atmospheric Sciences, University of Miami, Miami, FL, USA
[2] El Colegio de la Frontera Sur. Departamento de Ecología Acuática, Chetumal, Q. Roo, México.

Studies on biogeography of stony corals from the eastern Pacific have been conducted in detail only for reef species, and to date there have been no attempts to explain the differences of regional species richness on the basis of oceanographic conditions. The objective of this work was to determine the relationship between deep-water (> 200 m) scleractinian species richness along the western coast of the Americas, and four oceanographic variables (temperature, nitrates, phosphates and silicates), and the feasibility to use this information to model effects of global warming on those associations. Data on coral distribution were gathered from bibliography and museum collections, while information on oceanographic conditions from 200 to 2,000 m depth was obtained from NOAA atlases. Species richness, estimated for intervals of 5 degrees of latitude, was correlated with abiotic factors using the Spearman rank coefficient. In the Northern Hemisphere, total species richness was positively influenced by temperature, but negatively by nutrients. In contrast, there was no effect of those factors on coral diversity in the Southern Hemisphere. At the family level, high temperatures favored species richness of Caryophylliidae and Dendrophylliidae north of the Equator, but diversity was reduced in areas of high concentration of silicates. In the Southern Hemisphere, temperature was not associated with deep-water coral richness, but correlated negatively with diversity of Caryophylliidae. Nutrients also showed an inverse relationship with richness of the latter family. In the rest of the families analyzed there was no apparent effect of oceanic conditions on species richness in the Southern Hemisphere. The results indicated that richness may be influenced by changes in oceanographic factors (especially temperature and silicate concentration). Then, it is feasible to develop numerical models to predict possible changes in deep-water coral diversity on the basis of scenarios from global warming models.

A NEW LABORATORY METHOD FOR MONITORING DEEP-WATER CORAL POLYP BEHAVIOUR

J.M. Roberts and R.M. Anderson
Scottish Association for Marine Science, Dunstaffnage Marine Laboratory, Oban, Argyll, UK

Deep-water corals are found along the oceanic margins world-wide and in the north east Atlantic the most abundant species is *Lophelia pertusa* (L.). There is now growing evidence that deep-water reefs formed by such species are coming under increasing pressures from resource exploitation, principally deep-sea trawling and hydrocarbon exploration. Here a novel and unobtrusive method of recording deep-water coral behaviour in the laboratory is described using time-lapse video to record silhouettes of the polyps under infrared illumination. The polyps of *L. pertusa* behaved asynchronously and did not show any clear diurnal patterns over a three-day observation period. Conceptually, sessile benthic suspension feeders appear to be vulnerable to smothering by sediments disturbed by bottom trawls or sub-seabed drilling. This method allows deep-water coral polyp behaviour to be continuously monitored in the laboratory and therefore the responses of coral polyps to environmental perturbations such as sedimentation can be recorded. Further work is necessary to resolve the sensitivity of deep-water corals to short-term environmental change and the combined approach of *in situ* monitoring and subsequent laboratory experimentation has great potential to address these issues.

OBSERVATIONS OF *LOPHELIA PERTUSA* AND THE SURFICIAL GEOLOGY AT A DEEP-WATER SITE IN THE NORTHEASTERN GULF OF MEXICO

W. W. Schroeder
Marine Science Program, University of Alabama, and Dauphin Island Sea Lab, Dauphin Island, AL, USA

Authigenic carbonate, precipitated in conjunction with biogeochemical activity associated with hydrocarbon and related fluid seepage, provides exposed and buried hard substrate on the crest and flanks of a low-relief mound located on the upper De Soto Slope in the northeastern Gulf of Mexico. *Lophelia pertusa* has successfully colonized some of this carbonate material. Individual colonies range in size from a few centimeters to over 1.5 m in diameter while aggregations of closely associated colonies with linear orientations were observed to attain 1.5-2 m in height and width and 3-4 m in length. Many of the aggregated colonies appear to be in the first phase of the "thicket" building stage described by Squires (1964). Colonies less than 50-75 cm in diameter were nearly always completely pure white. Larger colonies and the aggregated colonies are often light to dark brown in coloration at their base and center with many having only white terminal branches and some with no white corallum at all.

PATTERNS OF ISOTOPIC DISEQUILIBRIA IN AZOOXANTHELLATE CORAL SKELETONS

Jodie E. Smith, Henry P. Schwarcz and Michael J. Risk
School of Geography and Geology
McMaster University, Hamilton, Ontario, Canada

A specimen of *Desmophyllum cristagalli*, an azooxanthellate (non-photosynthetic) coral was intensively and systematically sampled on all interior and exterior surfaces. Even though the coral grew at an almost constant temperature of about 2.5°C, XX^{18}O varied by almost 3% and was up to 3% depleted with respect to aragonite-seawater oxygen isotope equilibrium. Contour maps of XX^{18}O show that, although portions of the skeleton approached equilibrium, the locations of those areas were unpredictable and were not associated with any readily identifiable characteristics, such as colour, texture or crystalline structure changes. The use of regression lines and intercepts will give the mean temperature experienced by individual corals (Smith et al., 2000), but the prospect of documenting temperature changes over the lifetime of an individual coral remains problematical. Because of the large and seemingly random degree of isotopic disequilibrium, several isotopic values for coeval skeletal material must be obtained for the determination of a single temperature. Although azooxanthellate corals have been shown to have growth banding, analogous to reef corals, the layers are thin and difficult to see with the naked eye, uneven and often discontinuous, rending sampling for a 'time series' impossible at present. Reasons for the degree of variation in the isotopic patterns remain unclear.

AXIAL ROD GROWTH AND AGE ESTIMATION OF THE SEA PEN, *HALIPTERIS WILLEMOESI* KÖLLIKER

Matthew T. Wilson[1], Allen H. Andrews[2], Annette L. Brown[1], and Erik E. Cordes[3]

[1] National Marine Fisheries Service, Alaska Fisheries Science Center, Seattle, WA, USA

[2] Moss Landing Marine Laboratories, Moss Landing, CA, USA

[3] Biology Department, Pennsylvania State University, University Park, PA, USA

Halipteris willemoesi is a large octocoral commonly found in the Bering Sea. It is a member of a ubiquitous group of benthic cnidarians called sea pens (Octocorallia: Pennatulacea). Sea pens have a skeletal structure, the axial rod, that in cross section exhibits growth rings. Pairs of adjacent rings, or ring couplets, were assumed to be annuli and were used to estimate the age and growth of *H. willemoesi*. Twelve axial rods, extracted from *H. willemoesi* collected in the Bering Sea, were selected to represent small (25 - 29 cm total length), medium (97 - 130 cm TL) and large (152 - 167 cm TL) colonies. Each rod resembled a tapered dowel; the thickest part (0.90 - 6.75 mm in diameter) was at about 5-10% of total length from the base tip, the distal part was more gradually tapered than was the base. The number of ring couplets increased with rod size indicating their utility in estimating age and growth. Estimated age among rods was based on couplet counts at the thickest part of each rod; the average estimated age (±SE) was 7.1 ±0.7, 19.3 ±0.5, and 44.3 ±2.0 yr for small, medium, and large-size rods, respectively. Based on these estimated ages, average growth rate in total length was 3.9 ± 0.2, 6.1 ± 0.3, and 3.6 ± 0.1 cm yr^{-1} for small, medium, and large-size colonies. The average annual increase in maximum rod diameter among all colonies was 0.145 ± 0.003 SE mm yr^{-1}; therefore age prediction from maximum rod diameter was calculated (estimated age = 7.0(maximum rod diameter) - 0.2; $R^2 = 0.99$). At maximum diameter the average couplet width was relatively constant among the three colony sizes (0.072 ± 0.05 mm). X-ray diffraction and electron microprobe analyses revealed that the inorganic portion of the rod is composed of a high-magnesium calcite. Radiometric validation of these age and growth rate estimates was attempted, but high amounts of exogenous ^{210}Pb precluded using the disequilibria of ^{210}Pb:^{226}Ra. Instead, ^{210}Pb activities were measured in a series of cores extracted along the axial rod. These activities ranged from 0.691 ± 0.036 (SE) to 2.76 ± 0.13 dpm g^{-1}, but there was no pattern of decay along the length of the rod; therefore, the growth rates and corresponding ages could not be validated. Based on estimated age from ring couplet counts, growth in total rod length is slow at first, fastest at medium size, and slows toward maximum size, with an estimated longevity approaching 50 yr.

ABSTRACTS OF ORAL PRESENTATIONS

REPRODUCTION AND LARVAL BIOLOGY OF THE IVORY TREE CORAL, *OCULINA VARICOSA*

Sandra D. Brooke
Harbor Branch Oceanographic Institution, Division of Marine Science, Fort Pierce, FL, USA

The Ivory Tree Coral, *Oculina varicosa,* forms extensive reefs of azooxanthellate colonies along the edge of the Florida Hatteras shelf, at depths of 70-100 m. Healthy reefs support invertebrate and fish communities as diverse as those of tropical coral reefs; however, large sections of the reefs have been badly damaged. In an effort to protect the *Oculina* habitat, an area of 300 square miles is currently closed to bottom fishing, and a small-scale project was initiated in 1997 to restore some of the damaged area. An understanding of life history is essential for managers to develop effective restoration plans, but little is known about the reproduction or life cycle of this species. For most deep-sea corals, sampling is limited by logistics and cost. In the case of *O. varicosa*, conspecific colonies exist as smaller zooxanthellate forms on shallow (6 m) limestone ledges. I used these shallow colonies as a model to elucidate the reproductive strategy of this species. *Oculina varicosa* is a broadcast spawning species, with colonies of separate sexes. The gametogenic cycle begins in the early summer and spawning occurs during July and August, with no obvious relationship to lunar phase. Planulae show strong negative geotaxis for the first 12 hours of larval life, after which they sink and begin benthic probing. Settlement was observed after 20 days in culture. In 1999, the deep reefs were sampled to determine habitat-specific differences in reproduction of *O. varicosa*. Fecundity of the deep colonies was significantly lower than shallow colonies. Gametogenesis is concurrent in the two populations.

REPRODUCTION AND GROWTH OF *ANTHOMASTUS RITTERI* (OCTOCORALLIA: ALCYONACEA) FROM MONTEREY BAY, CA, USA

Erik E. Cordes[1], Dr. James Nybakken[1], Gilbert VanDykhuizen[2]
[1] Moss Landing Marine Labs, Moss Landing, CA, USA
[2] Monterey Bay Aquarium, Monterey, CA, USA

Anthomastus ritteri Nutting, 1909 is a deep-sea alcyonacean from the Pacific coast of California, U.S.A. and Baja California, Mexico. While descriptive studies exist, detailed information on the reproduction and life history of any species of *Anthomastus*, and deep-sea octocorals in general, is lacking. Laboratory specimens of *A. ritteri* were studied with respect to their reproductive biology, timing of larval settlement, and growth rates. Collected *A. ritteri* colonies were dioecious larval brooders that exhibited continuous reproduction. Developing larvae were brooded in the siphonozooids of the colony. The demersal planula larvae were capable of settlement 2 days after release.

The longest competency period recorded in the laboratory was 123 days. A Gompertz growth model was based on the size-specific growth rates of 15 colonies. Growth was slow for the first 1-2 years, but increased when colonies were comprised of 3-4 feeding polyps. Growth slowed in adult colonies, approaching an asymptotic size. In the model, *A. ritteri* approached asymptotic size between 25 and 30 years of age. This age is greater than those reported for tropical gorgonian colonies. These results agree with the general notion that growth rates are reduced and longevity increased in deep-sea species.

TASMANIAN SEAMOUNTS MARINE RESERVE

Nancy Dahl-Tacconi
Scientific Officer, Marine Protected Areas Environment Australia, Department of Environment and Heritage, Commonwealth of Australia

Australia's first exclusively deep-sea marine reserve, located 170 kilometres south of Hobart in the Southern Ocean, was declared in May 1999. The Tasmanian Seamounts Marine Reserve will add a unique sample of deep-sea features to the National Representative System of Marine Protected Areas in Australian waters. The Reserve was designed to protect unique and vulnerable benthic ecosystems within an area of approximately 370 km^2.

The seamount region off Tasmania is a distinct geological feature not found elsewhere on the continental margin of Australia. It includes 70 submerged and extinct volcanoes in water between 1000 and 2000 metres deep on the continental slope and supports an incredibly diverse range of benthic plants and animals, many of which are new to science and have not been found anywhere else in the world. Approximately 20% of this region, which has not been trawled and is in pristine condition, is included within the Reserve.

The Reserve comprises a vertical zoning system, in which the area below a depth of 500 m is a 'highly protected zone' (no fishing and no mining or petroleum exploration) and the area above a depth of 500 m is a 'managed resource zone' (some pelagic fishing allowed). The plan of management for the Tasmanian Seamounts Marine Reserve is currently being prepared. The remoteness and unusual features of this reserve present interesting challenges for practical planning and management. The management plan will be developed through consultations with stakeholders; innovative and cooperative options for monitoring and enforcement will be explored.

DEEP WATER CORAL BANKS IN PORCUPINE BASIN (SOUTH WEST OF IRELAND, NORTH EAST ATLANTIC)

Ben De Mol[1], Ann Vanreusel[2], Jean Pierre Henriet[1], Rudy Swennen[3], Michael Ivanov[4] and Peter Croker[5]
[1] RCMG, University of Ghent, Ghent, Belgium
[2] Marine Biology, University of Ghent, Ghent, Belgium
[3] Afdeling Fysico-chemische Geologie, Haverlee, Belgium

[4] UNESCO-MSU Centre for Marine Geoscience, Moscow State University, Moscow, Russia
[5] Petroleum Affairs Division, Department of Marine and Natural Resources, Dublin, Ireland

While Porcupine Basin was already famous for the evidence of *Lophelia pertusa* in water depths ranging from 500 to 1100 m (Le Danois, 1948; Thomson, 1873), recent seismic surveys (Henriet et al., 1998; Hovland et al., 1994) have highlighted a much wider extension and variety in coral bank habitats in this basin.

By now, it is possible to identify three major provinces of coral banks occurrence in Porcupine Basin: (1) high-relief surface mounds which have a dimension of 5 by 1 km and a height of 200 m, (2) flanked to the north by a swarm (1200 km^2) of buried mounds (bioherms?), somewhat smaller and funny shaped, while (3) outcropping or buried conical, 150 m high mounds (single or in elongated clusters) occur on the eastern slope.

As far as can be inferred from shallow cores, the surface lithology predominantly consists of terrigenous silty clay with intercalation of biogenic rubble. The mounds contain a remarkable number of specimens of living and dead *Lophelia pertusa* and *Madrepora oculata*. Further faunal elements include Eunicidea and Polychaeta, Cirripedia, Cidaroidea, Munidea and Gastropoda. The living and dead assemblages were underlain by a significant layer of coral debris in a marly matrix.

Two genetic hypotheses associate coral bank formation with hydrocarbon seeps along faults, or the decay of gas hydrates. Nevertheless the growth of these coral banks, in recent geological times, has most likely been controlled by a favourable oceanic circulation, currents, and changes in water-masses and nutrient supply to the corals and associated fauna.

THE OCEANOGRAPHIC BOUNDARY CONDITIONS OF DEEP-WATER CORAL BUILDUPS ALONG THE NW EUROPEAN CONTINENTAL MARGIN

André Freiwald
Institut für Geologie und Paläontologie, Universität Tübingen, Tübingen, Germany

This contribution focuses on the pseudocolonial *Lophelia pertusa* in the NE Atlantic. With more than 800 identified species, these coral buildups represent outstanding 'oases' of rich life in the cold and predominantly deep-water environments. A critical factor which determines the large-scale distribution of deep-water coral buildups along the NW European continental margin is the depth of the shelf break. Broad shelf ramps with shallow shelf break locations (<250 m water depth) foster the formation of effective oceanographic frontal systems — the shelf break fronts — which separate the lesser saline coastal or shelf waters from the more saline oceanic waters. Deep-seated shelf breaks have no such frontal system. In this oceanographic configuration the more dense oceanic waters underflow the shelf waters thus enabling the oceanic benthic (and pelagic) ecosystems to prograde onto the shelf environment. This latter situation pertains over most areas off the mid-Norwegian Shelf. A correlation between the occurrence of *Lophelia* and the physical properties of water masses — mostly expressed in terms of temperature — was suspected already in the 1940s. Modelling oceanographic parameters (temperature, salinity and dissolved oxygen) which are aquired online from the U.S. National Oceanographic Data Center outlines an even stronger correlation towards waters with highest salinities and lowest oxygen concentrations in any slope or shelf area analyzed along the NW European continental margin.

CORAL HABITAT MAPPING AND SUBMERSIBLE DIVES ON THE SULA RIDGE REEF, NORWEGIAN SHELF

André Freiwald[1] and Veit Hühnerbach[2]
[1] Institut für Geologie und Paläontologie, Universität Tübingen, Germany
[2] Southampton Oceanography Centre, Empress Dock, Southampton, U.K.

The major scientific target of a cruise carried out in summer 1999 was to retrieve high-quality sponge samples from the seafloor for the sensitive detection and subsequent identification of marine natural compounds. In order to offer an economic method for the search of sponge-rich habitats within the very complex coral reef topography and to avoid damaging the benthic communities, the following strategy was chosen:

1. Echosounding and side-scan-sonar mapping of a defined seafloor grid on Sula Ridge.
2. Identification of sponge-rich target areas on the basis of the seafloor mapping surveys.
3. Dives with the manned research submersible JAGO for groundtruthing and sampling.

On the Sula Ridge, the bathymetric range for the *Lophelia* reefs is from 330-233 m water depth. The reefs are concentrated along the summit near the escarpment of the asymmetric Sula Ridge. The length of the main *Lophelia* reef complex on the Sula Ridge is approximately 13 km. The reefs on the Sula Ridge are 10-20 m high on average.

The sonographs which were acquired with the deep-towed sidescan sonar along the strike line of the ridge allow the recognition of numerous individual 1-2 km long and 100-200 m wide reefs which often are arranged *en echelon*. Measurements carried out with the submersible confirmed the distinctive offset of 20° in the orientation of the reefs in relation to the northeast-southwest strike of the Sula Ridge. The corals can be seen to grow on the ridge and furrow systems which were produced by grounding icebergs.

BLACK CORAL: HISTORY OF A SUSTAINABLE FISHERY IN HAWAII

Richard W. Grigg
Department of Oceanography, University of Hawaii, Honolulu, Hawaii, USA

A small but sustainable black coral fishery has existed in Hawaii since the discovery of the resource off Maui in 1958. The raw material is used to manufacture precious coral jewelry, an industry valued at about $25 million in Hawaii today. Black coral is the official gem of the State of Hawaii, so designated by the Governor in 1987. During the early years of the fishery, biological research produced estimates of growth rate, recruitment and mortality rates and patterns of distribution and abundance, all of which led to the development of an optimal size limit and annual maximum sustained yields for the major species. Compliance with these measures by the black-coral divers who use SCUBA and traditional methods of harvest, has produced a record of sustained yield as well as undiminished recruitment and growth over the history of the fishery. A survey completed in 1998 showed that 97% of the major bed consisted entirely of new recruitment of colonies below the recommended size limit. In spite of this record of stability within the fishery, new technology and the potential for increases in demand both locally and in the world market, pose new pressures on the resource and may require more stringent management measures such as stronger enforcement and limited entry.

DEEP REEF COMMUNITIES: TORTUGAS BANKS AND PULLEY RIDGE

Walter C. Jaap[1], David Mallinson[2], Albert Hine[2], Pam Muller[2], Brett Jarrett[2], and Jennifer Wheaton[1]
[1] Florida Marine Research Institute, St. Petersburg, FL, USA
[2] Department of Marine Science, University of South Florida, St. Petersburg, FL, USA

Large deep-reef banks are located south and west of the Dry Tortugas National Park (DTNP), Florida. Tortugas Bank is the largest: 24°38'N, 83°04'W (20 x 10 km, 25 to 45 m deep). Coral cover in some locations exceeds 40%. *Montastraea cavernosa* is the principal framework coral, occurring in several colour morphs: green, blue, and red. Colonies have a pancake, cone, or mushroom morphology. The fauna includes at least forty species of Scleractinia and black coral, Antipatharia are found in deeper portions of the Bank. Tortugas Bank began as a shallow-water reef at the beginning of this interglacial period (~9,630 to 10,670 years before the present). Unlike the reefs in DTNP, Tortugas Bank did not keep pace with Holocene sea-level change. We speculate deep-reefs did not keep pace with rising sea-level because: a. that there was a rapid increase in rising sea-level, b. elevated nutrients in the water column, or c. severe cold climate events impacted the area.

Pulley Ridge is a linear feature (possibly a paleo-shoreline) NW of DTNP

(25°30'N, 83°20'W, 70 m deep). Substrate is low-relief karst limestone and the surface is veneered with thin plates of *Agaricia* spp. Most probably, the species composition includes: *Agaricia lamarcki, A. fragilis,* and *A. undata*; *Montastraea cavernosa* is also common. Antipatharia, comatulid crinoids, haxactinella sponges, deep-water octocorals, and the green alga, *Anadyomene menziesii,* are also conspicuous on Pulley Ridge.

The study of these deep reefs was conducted using mixed-gas scuba, an ROV, and a one-man submersible.

DETAILED MAPPING OF A SHALLOW OCCURRENCE OF *LOPHELIA PERTUSA* (L.) IN THE KOSTERFJORD, SWEDEN

Lisbeth Jonsson and Tomas Lundalv
Tjaernoe Marine Biological Laboratory, Stroemstad, Sweden

The bank-forming cold water coral *Lophelia pertusa* is usually found at depths between 200 and 1000 m, but in some areas the 'fjord effect' makes it possible for the species to thrive in more shallow waters. This is the case in the Kosterfjord, at the northern part of the Swedish west coast, where *L. pertusa* is found at about 80 m. Numerous separate colonies of different sizes grow here in habitats with strong currents. Multi-beam echosounding and side-scanning sonar surveys were used to obtain detailed bathymetric data. Structures on the sea floor, fitting the requirements of *Lophelia* colonies, were located and ROV-techniques (Remotely Operated Vehicle) used to obtain ground-truth data.

Detailed mapping of the located colonies was conducted using dGPS, an underwater positioning system (Simrad HPR 410P), and software to plot the movements of the ROVs. Position, size, shape and health status (visually evaluated) were recorded for each colony.

Areas of dead coral and records from earlier surveys suggest that the distribution of *L. pertusa* was once considerably larger in the area. One reason for the decline may be the extensive shrimp trawling in the fjord. Stress, caused by particles from resuspension of bottom sediments during trawling, is suspected of causing a negative energy budget for the coral and slowly suffocating the polyps. Frequent tracks from trawl doors have been recorded with side-scanning sonar, some very close to live corals.

SEAMOUNT CORAL REEFS OF TASMANIA: DIVERSITY, ENDEMISM, IMPACTS OF TRAWLING AND CONSERVATION

J. Anthony Koslow and K. Gowlett-Holmes
CSIRO Marine Research, Hobart, Tasmania, Australia

Orange roughy are trawled on small seamounts peaking at 500-1000 m depth on the continental slope south of Tasmania. In 1995, a Marine Protected Area (MPA) was proposed for a group of 12 deeper seamounts (peaks >1150 m depth) to protect the hitherto undescribed benthic fauna. In 1998, the seamount benthic fauna was surveyed to assess its diversity and depth distribution, the impact of trawling, and the conservation value of the proposed MPA.

The invertebrate fauna on unfished, relatively shallow seamounts (peaks <1400 m depth) was dense, diverse and dominated by suspension feeders, in particular a reef-forming colonial hard coral, *Solenosmilia variabilis,* as well as a variety of hard and soft corals, hydroids, sponges and suspension-feeding ophiuroids and sea-stars. Two hundred and sixty two species of invertebrates and 37 species of fish were enumerated from the 11-day cruise, compared with 598 species of invertebrates reported from seamounts worldwide since the *Challenger* Expedition of 1872-76 (Wilson and Kaufmann 1987). Twenty four to forty three percent of the invertebrate species were new to science, and between 16 and 33% appeared to be restricted to the seamount environment.

A photographic survey showed that trawl operations effectively removed the reef aggregate from heavily fished seamounts. However, most species had a broad depth distribution, and the fauna unique to the region's seamounts appeared to be adequately represented within the MPA, which is now permanently established.

ECOLOGY AND GROWTH OF ANTARCTIC SCLERACTINIA

Lucie Marquardt
Alfred-Wegener-Institut für Polar- und Meeresforschung,
Columbusstrasse, Bremerhaven, Germany

The present study on Antarctic Scleractinia covers several topics (taxonomy, distribution, ultrastructure, growth). Samples were collected during the EASIZ II cruise onboard RV *Polarstern* in 1998 to the eastern Weddell Sea and the Drake Passage. Seven species were found: *Flabellum impensum, Flabellum flexuosum, Flabellum curvatum, Javania antarctica, Gardineria Antarctica, Caryophyllia antarctica* and *Fungiacyathus marenzelleri.* They are all ahermatypic and azooxanthellate. *Flabellum curvatum* was found for the first time in the Antarctic zone. Geographic and bathymetric distribution of these species seems to be affected by food supply and substrate type. Principally all seven species are found sympatrically on the eastern Weddell Sea Shelf. Around King George Island and in the Drake Passage only *Flabellum impensum, Gardineria antarctica* and *Fungiacyathus* were found. Many scleractinian skeletons are substrate for other taxa such as bryozoans and serpulids. Skeleton size of the species is in the range given by the taxonomic literature. Organic content in term of ash free dry weight is low, ranging between 6 and 12% of wet weight. *Flabellum impensum* and *Javania antarctica* show a skeletal microstructure which leads to stability.

The skeletons of the scleractinian corals show very typical growth patterns. Most species have visible crests and troughs on the theca, which occur irregulary, whereas the cross-cut of *Javania antarctica* shows typical tree-like growth rings. Stable isotope analysis ($\delta^{13}C$ and $\delta^{18}O$) indicated seasonality of growth in *Javania antarctica, Flabellum impensum* and *Caryophyllia antarctica,* but the data were not sufficiently conclusive to estimate the age of the investigated individuals.

ON THE ROLE OF THE FINE FRACTION OF SESTON IN THE DIET OF THREE ANTARCTIC OCTOCORALS: ECOLOGICAL IMPLICATIONS

Covadonga Orejas[1] and Josep Maria Gili[2]
[1] Alfred Wegener Institute for Polar and Marine Research, Columbusstrasse, Bremerhaven, Germany
[2] Institut de Ciències del Mar, Barcelona, Spain

Octocorals are a widely distributed and abundant group in Antarctic waters. Despite their role in temperate and tropical regions knowledge of the ecology of this group is still very scarce in polar ecosystems. One of the aspects recently studied in the frame of the EASIZ-SCAR Programme is their trophic role in benthic communities on the Weddell Sea shelf. For this purpose, feeding experiments were carried out with two species of Antarctic gorgonians: *Primnoisis antarctica* and *Primnoella* sp. Results show that seston components including the fine fraction of suspended organic matter (bacteria, organic particles, and phytoplankton) play an important role in the diet of these animals. It was long assumed that Antarctic ecosystems experience a long period of inactivity due to extended periods of ice cover every year. However, recent studies have presented evidence of a short period of feeding inactivity (only a few weeks) for some littoral Antarctic benthic suspension feeders. Parallel to these results it was shown that (mainly heterotrophic) nano- and picoplankton dominates water column communities almost the year round whereas (autotrophic) microplankton including algal blooms is of importance only in the summer months. If suspension feeders have developed adaptations to feed on small cells and at lower concentrations, the results obtained from these three gorgonians could be of high ecological significance. Such feeding adaptations would allow the suspension feeding organisms to feed, and hence remain active, for considerably longer periods of the year than previously thought.

MULTIBEAM SWATH BATHYMETRY MAPPING OF COLD-WATER CORAL REEFS ON THE MID-NORWEGIAN SHELF

Dag Ottesen[1], Terje Thorsnes[1], Leif Rise[1], Jan Helge Fosså[2], Pål Buhl Mortensen[2] and Kjell Olsen[3]
[1] Geological Survey of Norway, Trondheim, Norway
[2] Institute of Marine Research, Nordnes, Norway
[3] Norwegian Hydrographic Service, Stavanger, Norway

Numerous cold-water coral reefs occur on the Mid-Norwegian shelf. Our knowledge of these reefs is based on reports from fishermen, from mapping related to pipeline laying and from scientific investigations using ROV and other techniques. In order to increase our knowledge of the reefs, a pilot project involving multibeam swath bathymetry mapping of two areas was initiated in 1999 using SIMRAD EM1002. These are the Sula Reef, one of the world's largest known cold-water *Lophelia* reef complexes, and the Horseshoe Ridge.

The Sula Reef is located at 64°N, 8°E, on the Mid-Norwegian continental shelf, 50 km off the Norwegian coast at about 300 m water depth. The reef is 13 km long and up to 400 m wide. The multibeam mapping shows that the reef comprises about 500 individual coral mounds which, in places, have grown together to form a ridge. Within the reef complex, mound density varies between 20 and 200 coral mounds per km^2. The individual coral mound has a diameter in the order of 50 to 100 m at its base, and reaches up to 30 m in height. The reef is located along the northern flank of the Suladeep, a glacial trough eroded in Upper Jurassic claystones. The major part of the reef complex is situated on a ridge of Lower Cretaceous sedimentary rocks, more resistant to glacial erosion. Extensive glacial lineaments, primarily megaflutes formed in till which covers the bedrock ridge, show that the main ice-movement direction in the area was towards west-southwest. Several of these glacial lineations cross the main reef complex, and indicate that the location of the reef is not controlled by glacial surface forms on a regional scale, even though local control is evident. We believe that the major reef developed on the ridge primarily due to a combination of suitable current conditions and hard substrate.

The Horseshoe Ridge is a 100 m high, spectacular, horseshoe-shaped, arcuate moraine ridge situated at 300 m water depth (66°N, 11°E). No cold-water reefs have been reported from this area previously, but the multibeam data clearly indicate the existence of large cold-water reefs. These structures are found on the flanks of the moraine ridge, with a size comparable to the mounds found along the Sula Reef. No potential reef structures have been found on the main moraine ridge or on the flat seafloor surrounding the Horseshoe Ridge. ROV investigations are needed to confirm that the potential reef structures indeed are reefs. If confirmed, this case study would point to a close connection between seabed conditions (moraine structures formed more than 12 000 years ago) and the present-day distribution of cold-water reefs.

DIVERSITY AND ENDEMISM OF THE BENTHIC SEAMOUNT FAUNA IN THE SOUTHWEST PACIFIC

Bertrand Richer de Forges[1], J. Anthony Koslow[2] and G.C.B. Poore[3]
[1] Centre IRD de Noumea, Noumea Cedex, New Caledonia
[2] CSIRO Marine Research, Hobart, Tasmania, Australia
[3] Museum of Victoria, Abbotsford, Victoria, Australia

In their review of the seamount biota, Wilson and Kaufmann (1987) reported a total of 597 invertebrate species recorded from ~59 of the estimated 2 x104 seamounts in the world's oceans. Most reports, based on a single taxonomic group, were extremely limited—five seamounts accounted for 72% of the species recorded—but the review concluded that seamounts contain only a moderately diverse epibenthic fauna, of which 15% not known from elsewhere were potential seamount endemics.

More than 850 macro- and megafaunal species are now recorded from seamounts in the Tasman Sea and southeast Coral Sea, apparently only a fraction of the available species. Approximately 29 to 34% of these species are new to science and are potential seamount endemics. Species distributions appear to be highly localized. Median species overlap decreased from 21% for seamounts on the same ridge system to only 4% for seamounts at the same latitude but on ridge systems separated by ~1000 km. Not a single species from the Tasmanian seamounts occurred on the seamounts in the northern Tasman Sea. For comparison, ~60% of decapod crustaceans from soft sediments on the continental slope off SE Australia are recorded from the Indo-Pacific. These high levels of endemism have profound implications for the conservation of seamount faunas world wide.

LIFESPANS AND GROWTH PATTERNS OF TWO DEEP-SEA CORALS: *PRIMNOA RESEDAEFORMIS* AND *DESMOPHYLLUM CRISTAGALLI*

M.J. Risk[1], J.M. Heikoop[2] and R. Beukens[3]

[1] School of Geography and Geology, McMaster University, Hamilton, ON, Canada

[2] Los Alamos National Laboratory, Los Alamos, New Mexico, USA

[3] Department of Physics, University of Toronto, Toronto, ON, Canada

A subfossil fragment of the deep-sea gorgonian coral *Primnoa resedaeformis* was 14C AMS dated along a radial growth transect. Dates ranged from 2600 +/- 50 at the outside to 2920 +/- 60 ^{14}C years BP near the interior, an age of >300 years. The average radial growth was approximately 0.044 mm per yr. Based on comparison with live-collected specimens, we estimate the entire colony may have been about 0.5–0.75 m tall, suggesting a linear tip extension rate of 1.5-2.5 mm per year.

Towards the centre of the main stem, the coral skeleton is composed of alternating couplets, 200 microns in width, of gorgonin (a horn-like organic skeletal protein) and calcite. We believe these couplets are annual. Within this larger scale of banding are finer couplets of gorgonin and calcite, whose frequencies suggest lunar monthly periodicities. Both scales of banding may reflect periodicity in food supply from sinking POM. Outer skeletal growth is predominantly massive calcite with intermittent gorgonin layers. If precipitation was continuous, approximately 25 microns of radial growth would be deposited each year.

The scleractinian *Desmophyllum cristagalli* lives to >200 years, and has rates of linear extension of 0.5-1.0 mm per year. The skeletons show growth bands of approx. 10 microns width, which may be annual. Due to tissue extension and retraction in life, parts of the skeleton may be overgrown or suffer dissolution. Although we have shown in previous publications that ocean temperatures can be obtained from analysis of this coral, long annual records will be difficult to obtain.

ENVIRONMENTAL RECORDING – THE *LOPHELIA PERTUSA* MODEL

J.M. Roberts[1], S.R. Chenery[2], B. Spiro[2] and J.D. Gage[1]
[1] Scottish Association for Marine Science, Oban, Argyll, UK
[2] British Geological Survey, Kingsley Dunham Centre, Keyworth, Nottingham, UK

Cold-water azooxanthellate corals such as *Lophelia pertusa* are found from relatively shallow depths in the Norwegian fjords to greater depths of over 1 km on the continental shelf and offshore banks. Their aragonite skeleton may contain information about the past environment, including whether or not coral colonies have been exposed to discharges from hydrocarbon exploration. Given that *L. pertusa* is found over a wide geographic area we have analysed samples from a range of sites within the North East Atlantic. Analysis of carbon and oxygen stable isotopes has emphasised the importance of kinetic factors in azooxanthellate coral calcification. Trace element analysis has proved that it is possible to sample this material at very high spatial resolution and that a number of trace elements show some correlation with the kinetic factors driving the stable isotope composition. It seems likely that an important underlying environmental factor in controlling coral growth is the pulsed annual nature of surface water productivity during the spring bloom.

COLLECTION AND HUSBANDRY OF DEEP-WATER CORALS FROM MONTEREY BAY, CALIFORNIA

Edward W. Seidel
Husbandry Operations, Monterey Bay Aquarium, Monterey, CA, USA

In March of 1999 the Monterey Bay Aquarium opened the largest living exhibit of deep sea animals. Entitled "Mysteries of the Deep," this exhibit is the culmination of 8 years of intensive research into the collection, transport, husbandry, and long-term captive display of animals from Monterey's submarine canyon. In this talk we will cover our findings with the octocorals *Paragorgia, Euplexaura, Swiftia,* and *Anthomastus*, as well as the pennatulids *Halipteris, Anthoptilum,* and *Umbellula*.

POSTER ABSTRACTS

MANAGEMENT OF HAWAIIAN PRECIOUS CORALS USING MOLECULAR GENETIC METHODS

Amy R. Baco[1], Craig R. Smith[1] and George K. Roderick[2]
[1] Department of Oceanography, University of Hawaii, USA
[2] Environmental Science, Policy, and Management, University of California, Berkeley, CA, USA

Hawaiian deep-sea precious corals represent an extremely profitable fishery, yet little is known about the life history and dispersal of the exploited species. Several recent studies have provided evidence of significant genetic structure between shallow-water coral populations, including several species capable of long-distance dispersal. If significant genetic structure exists in populations of precious corals, this would suggest that the elimination (through overharvesting) of a bed of precious corals would result in loss of overall genetic diversity in the species. We are studying three species of Hawaiian deep-sea precious corals, to determine the appropriate management units (or "stocks") for each species. Stock identifications will be estimated using a recently developed microsatellite technique. Microsatellites are segments of DNA which consist of repeated units of short (di- or tri-nucleotides) sequences. Microsatellites are highly variable, making them ideal for kinship and population-level studies. By determining the stock structure of the harvested species and providing information on dispersal and recruitment in these species as a function of life history, this project will elucidate the effectiveness of the designated "Refugium" bed in the Northwest Hawaiian Islands for re-seeding exploited populations of precious corals. This information will substantially improve our ability to manage the Hawaiian coral fishery as a sustainable resource.

ARE THERE SPECIALIZED FISH SPECIES IN DEEP-SEA CORALS?

Nicolas Bailly
Institut de Systématique, Muséum National d'Histoire Naturelle, Laboratoire d'Ichtyologie Générale et Appliquée, Paris, France

Two years ago, I synonymyzed two nominal species of Macrouridae (Bailly, 1998). The first one described is known only by the holotype, the second one, by the holotype and three paratypes. The five specimens were caught in three trawls and were all associated with deep-sea coral (resp. in 1895, in the Bay of Biscay, 1410 m, and 1978-79 in the Porcupine Sea-Bight, 770-1042 m).

A review of the scarce literature will be presented relating the fish catches in this habitat to address this issue: could the trawl sampling depict specialized, maybe endemic, fish species in deep-sea corals?

The first results show that the other species caught by the same trawls are mainly bigger ones, as in the two cases above, which are also present in muddy habitats. As the deep-sea coral habitat seems to be patchy, they may have been caught outside the coral patches sampled as well. The trawl technology used to sample the ichthyofauna of these habitats should be abandoned for those used during the hydrothermal vents (habitat with which comparisons should be fruitful) surveys, i.e., with traps and submersible surveys, to catch small fishes possibly hidden in corals.

Generally, the rocky deep-sea ichthyofauna is poorly known and understood due to the sampling difficulties, as evidenced by rare catches of some species (e.g., *Gaidropsarus* sp. recently, Gadiformes).

Bailly, N. 1998. *Caelorinchus abditilux* Merrett, 1980, a junior synonym of *Caelorinchus caudani* (Koehler, 1896) (Teleostei: Macrouridae), J. Fish Biol 51(6): 1192-11.

HEXACTINELLID SPONGE REEFS ON THE BRITISH COLUMBIA SHELF: SUBMERSIBLE OBSERVATIONS AND GEOPHYSICAL MAPPING

K.W. Conway[1], M. Krautter[2], J.V. Barrie[1], W.C. Austin[3] and M. Neuweiler[2]
[1] Pacific Geoscience Centre, Geological Survey of Canada, Sidney, B.C., Canada
[2] University of Stuttgart, Germany
[3] Marine Ecology Station and Khoyatan Marine Lab

Globally unique sponge reefs dominated by species of hexactinellid sponges occur in the deep shelf troughs on the western Canadian continental shelf. Submersible dives reveal these reefs to consist of dense populations of hexactinosan sponges that cover bioconstructions that are up to 18 m high and kilometres wide. Acoustic mapping using sidescan sonar and high resolution seismic reveal a range of reef forms of both biostromal or biohermal character in water depths of between 180 and 250 m. The sponges trap suspended sediment as they baffle currents transporting sediment across the reef. The baffled sediment is organic-rich siliciclastic clay, transported in suspension as flocs by seafloor currents. Three species of hexactinosan sponge construct the skeletal framework through the biological fixing and deposition of opaline silica, which fuses the spicules of the skeleton of the individual sponge. This structure also allows inter- and conspecific attachment of young sponges onto skeletons of dead hexactinosan sponges, which permits the multigenerational habitation of reef sites. The mounds (bioherms) and sheet-like accumulations (biostromes) cover a low angle, non-depositional, iceberg-scoured seafloor, relict since the deglaciation of the region, about 13 thousand years ago. The bases of the oldest sponge reefs date from approximately 9000 years before present.

EARLY RECORDS OF DEEP-SEA CORALS FROM SUBMARINE TELEGRAPH CABLES RECOVERED OFF NOVA SCOTIA

Derek S. Davis, Andrew Hebda and Leslie Pezzack
Nova Scotia Museum of Natural History, Halifax, NS, Canada

Some of the earliest records of deep-sea corals from Eastern North America were made in the late 19th century and early 20th century during the recovery of submarine telegraph cables for repair. The Nova Scotia Museum of Natural History has in its collection specimens obtained by the Western Union Telegraph Company cable ship S.S. *Minia* between 1888 and 1913. Also, one of the ship's logs is now in the collection of the Maritime Museum of the Atlantic, Halifax, Nova Scotia. The coral samples were taken from the Cape Cod to Brest cable, at the mouth of the Gully off Sable Island and in the Northeast Channel between Browns Bank and Georges Bank. Samples were collected and records made of other invertebrates from the same cable southeast of St. Pierre Bank and north of Browns Bank, Nova Scotia. A single specimen of the scleractinian *Desmophyllum cristagalli*, collected off the Gully at a depth of about 1000 m, was attached directly to the cable and grew to a height of 4 cm in approximately 19 years. The gorgonian *Paragorgia arborea*, taken from the cable in the Northeast Channel at a depth of approximately 1000 m, showed extensive growth about 2 cm thick and extended from the cable in branches up to 20 cm in length after 44 years.

These types of records show the archival value of museum collections to current and future scientific research.

DISTRIBUTION AND CONCENTRATION OF FISH IN AREAS WITH DEEP-SEA CORALS

Dag Furevik[1], Leif Nøttestad[1], Jan Helge Fosså[1], Åse Husebø[1] and Sebastian Beck Jørgensen[2]
[1] Institute of Marine Research, Bergen, Norway
[2] University of Copenhagen, København, Denmark

Extensive longlining was conducted at Aktivneset (62°60'N, 3°50'E) on Storegga in southwestern Norway along the continental shelf with the commercial vessel M/V *Nyvoll Senior* during the period 17-25 June 1999 in order to map fish concentrations in relation to deep-sea corals. A comparative analysis was conducted during the period 3-4 June 1998 near Storegga at Sørmansneset with the longlining vessel M/V *Nesbakk*. Gillnet data from the vessel M/V *Anne Beate* was analysed and compared with longline catches in relation to concentrations of corals. A comparison was made to quantify differences between the amount and distribution of fish in areas containing deep-water coral reefs at depths between 150 and 300 m and areas outside these corals at similar depth ranges. In total 26 stubs of longline were investigated, each consisting of approximately 2600 hooks. Significantly larger catches of commercially important redfish (*Sebastes* sp.) on stubs placed above deep-water corals was found compared to stubs placed outside coral reefs. Ling and tusk had also larger catches above corals than outside, although not statistically significant. Redfish, tusk and ling caught along coral reefs were also generally larger in size than fish caught outside on softer bottom substrate. The most common fish species that were caught along corals at different depths were redfish, tusk and ling, while cod, saithe, angler-fish and Atlantic halibut were caught in small concentrations. The coral reef species that were caught occasionally on longline included *Primnoa placomus, Lophelia pertusa, Primnoa resedaeformis* in addition to *Phakalia ventilabrum*. The catches of fish species were clearly distributed in well-defined patches and concentrations of redfish, tusk and ling came often along with remnants of corals caught up on the hooks. This indicates that these fish species probably prefer habitats containing corals. Analysis of underwater video recordings from a remotely operated vehicle (ROV) supports this theory.

PRELIMINARY BIOLOGICAL OBSERVATIONS OF COLD AND DEEP-WATER CORAL ASSOCIATED WITH NEWLY DISCOVERED CARBONATE/MUD MOUNDS FOUND ALONG THE PORCUPINE AND ROCKALL MARGINS, OFF THE WEST COAST OF IRELAND

A.J. Grehan[1], H. de Haas[2] and T. Van Weering[2]
[1] Martin Ryan Institute, National University of Ireland, Galway, Ireland
[2] Netherlands Institute for Sea Research (NIOZ), Texel, The Netherlands

Geophysical prospection of the Porcupine and Rockall Margins off the west coast of Ireland has revealed numerous raised seafloor features that have been shown to be biogenic in origin. These enigmatic carbonate/mud mounds which are up to 300 m high are the subjects of intense research to discover the mechanism of their genesis, growth and maintenance. Many of the mounds support dense stands of living cold-water coral (*Lophelia pertusa* and *Madrepora aculata*) at depths between 600 and 1000 m. Here we present preliminary observations of the living cold-water coral ecosystem sampled from carbonate and mud mounds in the area, aboard the NIOZ RV *Pelagia* (Cruise 64 PE 143).

EVIDENCE FOR PAST TRAWLING DAMAGE TO COLD-WATER CORALS

D. Long[1] and J.M. Roberts[2]
[1] British Geological Survey, Edinburgh, UK
[2] Scottish Association for Marine Science, Oban, Argyll, UK

As part of the British Geological Survey's mapping programme, more than 3000 sediment samples have been taken in water depths of more than 200 m in the northeast Atlantic. Some of these samples contained fragments of cold-water corals. BGS sample site 59-07/293, located at 59°43.4'N, 6°28.9'W in 282 m of water at the eastern end of the Wyville-Thomson Ridge, recovered abundant fragments of *Lophelia pertusa* at various levels within the core, including fragments of *L. pertusa* that appeared to have been broken and buried while still alive.

At the base of surface sands (5 cm) two fragments of bryozoan-encrusted subrounded coral were found. Between 20 cm and 63 cm subseabed depth there are abundant fragments of fresh-looking coral, with angular fractures and mostly upside down. Fragments of coral occurred spasmodically down core to 3.25 m. The lower fragments are frequently much thinner than fragments near the top of the core. Below 3.25 m depth a stiff diamict of glacial age occurs. The variety of fragments within the single core 59-07/293VE suggests they are derived from separate sources and are of different ages. A dating programme is currently underway. The uppermost samples are interpreted as part of a mobile surface layer. Possible explanations of the emplacement of the coral-rich sequence between 20 cm and 63 cm subseabed depth include submarine landslides and deep-sea trawling. The latter is considered most likely to explain the freshness of the material. The lowermost fragments suggest that *L. pertusa* has grown in this area since the end of the last glaciation about 13,000 years ago.

OCCURRENCE OF *LOPHELIA PERTUSA* OFFSHORE THE UK

D. Long[1], J.M. Roberts[2] and E.J. Gillespie[1]
[1] British Geological Survey, Edinburgh, UK
[2] Scottish Association for Marine Science, Oban, Argyll, UK

Records of *Lophelia pertusa* in the northeast Atlantic date back more than 200 years. In the last twenty years parts of this region have been open to hydrocarbon exploration and are under increasing pressure from deep-water trawling. To aid reconciliation of conflicts of interest we have produced an updated map of *L. pertusa* distribution offshore the UK. This incorporates data from a wide range of sources. It shows that *L. pertusa* is concentrated on the continental slope and offshore banks, particularly in areas of low terrigenous input. The coral has been found attaching itself to rock platforms, dropstones and artifacts including shipwrecks and oil industry infrastructure. Its distribution is strongly controlled by the oceanographic regime. Within the Rockall Trough, *L. pertusa* is found to more than 1000 m depth but west of Shetland it is absent below 500 m depth where cold bottom waters (<4°C) occur. This distribution does not match that of enigmatic seabed mounds (typically 10-100 m diameter, 1-4 m high) noted on acoustic seabed surveys. These features have previously been interpreted as coral reefs but have now been shown to be sedimentary structures and only locally provide support for growths of *L. pertusa* where oceanographic conditions favour the growth of deep-water coral.

OCCURRENCE OF *LOPHELIA PERTUSA* ON NORTH SEA OIL RIGS

J.M. Roberts
Scottish Association for Marine Science, Oban, Argyll, UK

Until 1999 the only verified locations of *Lophelia pertusa* in the North Sea were of two sites where trawls had recovered dead coral. It is now becoming clear that structures around several oil platforms in the northern North Sea have been colonized by *L. pertusa*. The size of the colonies suggests that they may have recruited shortly after the structures were put in place. The colonies are restricted to depths below 60 m where there is the year-round presence of cooled Atlantic water. It seems likely that coral planulae will have been transported into the North Sea from the Atlantic margin via the East Shetland Atlantic Inflow current. Given that the North Sea seafloor is predominantly muddy, the presence of hard substrate in the form of steel and concrete platforms has provided an attachment site for a variety of suspension feeders including soft corals, anemones and hydroids. The distribution of conspicuous megafauna on a tanker mooring in the Beryl Alpha field is described.

LABORATORY STUDIES OF *LOPHELIA PERTUSA* – PRELIMINARY STUDIES OF POLYP BEHAVIOUR

J.M. Roberts and R. Anderson
Scottish Association for Marine Science, Oban, Argyll, UK

To date there have been few studies that have examined deep-water corals in the laboratory. Here we describe an aquarium system used to maintain *Lophelia pertusa* in controlled laboratory conditions. This allows study of coral behaviour and physiology. Coral polyp behaviour has been examined using time-lapse video under infra red illumination. The polyps have been shown to act independently of one another, spontaneously contracting and expanding without any clear external cue. To date there is no evidence for any diurnal pattern in this behaviour. A simple index of polyp behaviour has been used to examine the response of *Lophelia pertusa* to sand deposition. Sand deposition at 0.1 $mg.cm^{-2}.min^{-1}$ reduced the level of polyp extension but did not eliminate it altogether. Further work is planned to examine the sensitivity of this species to sedimentation at different rates and from different sources.

ENVIRONMENTAL MONITORING – PAST STUDIES AND FUTURE APPROACHES TO UNDERSTANDING DEEP-WATER CORAL OCCURRENCE

J.M. Roberts[1], A. Cunningham[2] and J.D. Gage[1]
[1] Scottish Association for Marine Science, Oban, Argyll, UK
[2] Applied Physics Department, University of Strathclyde, Glasgow, UK

In the past there have been two main hypotheses to explain why deep-water corals banks are found where they are. The first relates coral banks to areas of locally increased current or where internal tidal waves impinge on critical slopes. The second postulates that coral banks rely on a chemosynthetic food chain fuelled by light hydrocarbon seeps. The first hypothesis suggests that in those areas where currents are locally increased, coral growth is favoured by a more abundant food supply and passive sediment removal. Here we describe a preliminary investigation of the first of these two hypotheses. The environment around a coral bank off the Norwegian coast at Leksa was examined. A mooring deployment showed that this coral bank was subject to a strongly tidal current regime with peak current flows of up to 40 $cm.s^{-1}$. A data-logging optical instrument designed to measure phytopigment fluorescence, beam attenuation, and forward and backward light scattering revealed a suspended particle level of about 6-6.5 $mg.l^{-1}$. Future approaches to *in situ* monitoring of deep-water corals will be discussed.

NO MORE DREDGING IN THE DARK: TOWARDS LESS DAMAGING METHODS OF SAMPLING COLD-WATER CORALS

J.M. Roberts[1], P.B. Mortensen[2] and R.C. Sundt
[1] Scottish Association for Marine Science, Oban, Argyll, UK
[2] Institute of Marine Research, Nordnes, Bergen, Norway

Traditional techniques used to sample deep-water corals by dredge and trawl cause extensive impact to both the coral bank and surrounding seabed. However, these techniques are cheap, readily available and do not require a specialized vessel. Modern sampling techniques using submersibles and remotely operated vehicles (ROV) cause very little or no damage, but are expensive and require highly specialized vessels. Here we describe a combination of video camera and benthic grab for sampling deep-water corals and test this method on a *Lophelia pertusa* bank in Osterfjorden, western Norway. The video-assisted grab was successfully used both for locating and sampling *L. pertusa* and confirmed the presence of this species in a location last sampled over forty years ago. This method can largely replace the use of traditional, more destructive dredging and trawling techniques without recourse to expensive submersible-based systems.

INVERTEBRATE FAUNA ASSOCIATED WITH DEEP-SEA CORAL COMMUNITIES OFF NOVA SCOTIA

Heidi Schaefer
Independent researcher. Recipient of Rare Species Grant from the Nova Scotia Museum of Natural History, Halifax, NS, Canada

There is currently great interest in Nova Scotia's deep-sea gorgonian and scleractinian corals. Recent studies have examined deep-sea coral distribution and status, however important information gaps relating to deep-sea coral ecology still exist.

This study proposes to assemble a systematic list of those invertebrate species which are closely associated with Nova Scotia's deep-sea corals through the gathering together of scientific information from literature and from accessible museum collections and records. In addition to the list, species will be placed in ecological groupings (e.g., epizoic, hard substrate, soft substrate) and the ecological niche of species will be noted when known. The research area would comprise the Scotian Shelf and Georges Bank, from Georges Bank to the Laurentian Channel. The assembly of a complete and up-to-date list of invertebrate faunal species associated with deep-sea corals will allow for an evaluation of the Nova Scotia Museum of Natural History's collection of the listed species.

This research will enhance our understanding of deep-sea coral habitat and increase our knowledge of Nova Scotia's benthic marine biodiversity. As a result, we will be better able to understand and sustainably manage our marine ecosystems.

MARINE INVERTEBRATE DIVERSITY INITIATIVE

Heidi Schaefer, Jayne Roma and Lesley Carter
Ecology Action Centre, Halifax, Nova Scotia, Canada

The Marine Invertebrate Diversity Initiative (MIDI) is a project partnered by the Ecology Action Centre (EAC), Environment Canada, Fisheries and Oceans Canada, the Integrated Coastal Planning Project, Dalhousie University, the Nova Scotia Museum of Natural History, and a number of private citizens to produce a standard reference in the form of a Web-accessible database for marine invertebrates of the Bay of Fundy, Gulf of Maine and Scotian Shelf. Its ultimate goal is to increase our knowledge of marine invertebrate biodiversity, and thereby to engage scientists, educators, resource users, environmental managers, environmentalists, and the general public in learning, conserving and caring more about the biodiversity of our oceans.

This project was initiated first to remedy the lack of commonly accessible and comprehensive information on marine invertebrate biodiversity in eastern Canada and the Gulf of Maine, and second to incorporate knowledge of marine invertebrates fully into the management of human activities affecting marine habitats and resources.

The poster presentation will display the MIDI information system, a map of the study area, and photos of selected invertebrate species. We would also have computer and Internet access to display our website database, and to focus this presentation on habitats and fauna of primary interest; deep-sea corals, saltmarshes and mudflats. This selection is being made in support of ecosystem studies and conservation projects currently underway.

VIDEO DOCUMENTATION OF THE GEOLOGY AND DISTRIBUTION OF *LOPHELIA PROLIFERA* AT A DEEP-WATER REEF SITE IN THE NORTHEASTERN GULF OF MEXICO

William W. Schroeder
Marine Science Program, The University of Alabama, Dauphin Island, AL, USA

Video documentation from a towed camera survey and manned submersible dives are utilized to describe the geology and distribution of *Lophelia prolifera* at a deep-water reef site in the northeastern Gulf of Mexico.

The reef complex is situated on the crest and flanks of a low-relief mound located on the upper De Soto Slope at 29°09.5'N, 88°01.0'W. The mound has a relatively flat top, moderately sloping to steep sides and rises between 45 to 90 m above the surrounding sea floor to a minimum water depth of 428 m. Bottom sediments consist of unconsolidated clay and silty clay and authigenic carbonate deposits, in the form of crusts, outcrops and rubble.

The southern rim of the top of the mound has an overall moderate slope and consists of a combination of locally hummocky terrain made up of carbonate knolls and ridges, some with abrupt vertical relief, and relatively flat terraces built from carbonate substrate (large plates and slabs, moderate to small boulders, and small rubble), and unconsolidated sediments. Colonies of *L. prolifera*, up to one metre in height, cover many of the larger carbonate features. To the immediate east and southeast the bottom levels off into a relatively flat area consisting mostly of soft, unconsolidated sediment and patches of carbonate outcrops many with colonies of *L. prolifera*.

The slope on the western side of the mound is steep and consists of a series of ridges/hummocks and swales of unconsolidated fine sediment. These features are oriented along-isobath and have a vertical relief of 1 to 3 m from crest to trough. Carbonate outcrops occasionally occur on the ridges and support colonies of *L. prolifera*. At the maximum depth surveyed on the western flank (496 m) a furrow cut into the soft sediment was observed that extended out of sight both up- and down-slope. This disturbance in the bottom was traced back up-slope. The furrow changed from a clean, narrow (10 cm) cut into the soft sediment to a complete disruption of the surface (e.g., sediments overturned and either piled up into small ridges/mounds or scraped away forming trenches/depressions) over a 2 m swath. No extensive destruction of carbonate structures was recorded, however, numerous colonies of *L. prolifera* were observed to have been severely damaged. Based on these observationz the most likely cause of this disruption is surmised to be a wire cable that struck the bottom one or more times.

The upper portion of the southern and southwestern flank of the mound has a steep to very steep slope covered with fine, unconsolidated sediment and small, terrace-like features built of exposed carbonate outcrops covered with *L. prolifera.* The lower portion appears to be small, steep-sided valley which may have formed as a result of slumping. At the base of the "slump zone?" the bottom flattens out into at least one terrace-like feature covered with large, mostly irregular-shaped carbonate blocks (up to 3 to 4 m tall by 1 to 2 m wide) and slabs (4 to 5 m across). Most of the surface of this carbonate material is barren, however large isolated colonies of *L. prolifera* were observed at this site.

WORLD ATLAS OF CORAL REEFS

M.D. Spalding, E.P. Green and C. Ravilious
UNEP-World Conservation Monitoring Centre, Cambridge, UK

Since 1993, the UNEP World Conservation Monitoring Centre has been developing the most detailed global map of shallow-water coral reefs compiling coral reef features into a multi-national, multi-source ARCINFO GIS. In combination with related information on the biodiversity and status of coral reefs the maps are an important contribution to global efforts to better understand and protect these critical ecosystems at the global scale. The *World Atlas of Coral Reefs* will take this information to a wider audience providing a critical information resource to scientists, policy-makers and planners at national and international levels.

The *World Atlas of Coral Reefs* will provide a detailed presentation of the distribution and status of coral reefs around the world, presenting maps, tables and descriptive texts. Information on the geography, biodiversity, human use, and threats to the continuing existence of coral reefs will be presented, linked to nearly 100 detailed colour maps and fully referenced bibliographic sources.

UNEP-WCMC would like to increase the prominence of deep-sea corals in international conservation by including maps of these reefs in the atlas. However we are aware of only limited sources of data on the distribution of deep-sea coral reefs. Contributions to the atlas would therefore be most welcome.

LIST OF PARTICIPANTS AT THE FIRST INTERNATIONAL SYMPOSIUM ON DEEP SEA CORALS, JULY 30 – AUGUST 2, 2000, HALIFAX, NOVA SCOTIA, CANADA

Mitch Anderson, 131 Water St., Suite 214, Vancouver BC, V6B 4M3, Canada, manderson@sierralegal.org

Allen H. Andrews, Moss Landing Marine Laboratories, 8272 Moss Landing Rd., Moss Landing, CA, 95039-9647, USA, astrofish@redshift.com

Lora Annis, National Sea Products Limited, 100 Battery Pt., PO Box 910, Lunenburg, NS, BOJ 2C0, Canada, annis@highlinerfoods.com

Sanford Atwood, PO Box 97, Clarks Harbour, NS, B0W 1P0, Canada

Ivar G. Babb, National Undersea Research Center, UCONN at Avery Point, 1084 Shennecossett Road., Groton, CT 06340, USA, babb@uconnvm.uconn.edu

Amy Baco-Taylor, University of Hawaii, Dept. of Oceanography, 1000 Pope Road Honolulu, HI, 96822, USA, abaco@soest.hawaii.edu

Nicolas Bailly, Museum National d'Histoire Naturelle, Laboratoire d'Ichtyologie Général et Appliquée, 43 rue Cuvier, 75231 Paris cedex 05, France, bailly@mnhn.fr

Susan Belford, Jacques Whitford Environment Limited, 3 Spectacle Lake Drive, Dartmouth, NS, B3B 1W8 Canada, sbelford@jacqueswhittord.com

Maya Borel Best, National Museum of Natural History, PO Box 9517, 2300 RA Leiden, The Netherlands best@naturalis.nnm.nl

Cynthia Bourbonnais, Department of Fisheries and Oceans, Bedford Institute of Oceanography, Dartmouth, NS, B2Y 4A2, Canada, bourbonnaisc@mar.dfo-mpo.gc.ca

Sandra Brooke, Harbour Branch Oceanographic Institution, 5600 US1, North Fort Pierce, FL, 34946, USA, brooke@hboi.edu

Andrew Bruckner, National Marine Fisheries Service, 1315 East West Highway, Silver Spring, MD 20910, USA, andy.bruckner@noaa.gov

Mark Butler, Ecology Action Centre, 1568 Argyle St., Suite 31, Halifax, NS, B3J 2B3, Canada, ar427@chebucto.ns.ca

Maria-Ines Buzeta, Department of Fisheries and Oceans, Biological Station, 531 Brandy Cove Rd., Saint Andrews, NB, E5B 2L9, Canada, buzetam@mar.dfo-mpo.gc.ca

Stephen D. Cairns, Dept. of Invertebrate Zoology, National Museum of Natural History, Smithsonian Institution, Washington, DC, 20560-0163, USA, cairns.stephen@nmnh.si.edu

Beth Cameron, Dept. of Oceanography, Dalhousie University, 1355 Oxford St., Halifax, NS, B3H 4J1, Canada, beth.cameron@ns.sympatico.ca

Scott Coffen-Smout, Oceans Act Coordination Office, Fisheries and Oceans Canada, PO Box 1006, Mail Stn. B500, Dartmouth, NS, B2Y 4A2, Canada, coffen-smouts@mar.dfo-mpo.gc.ca

Kim Conway, NRCAN, Geological Survey of Canada, Pacific Geoscience Centre, PO Box 6000, Sidney, BC, V8C 4B2, Canada, conway@pgc.nrcan.gc.ca

Erik E. Cordes, Moss Landing Marine Labs, PO Box 450, Moss Landing, CA, 95039, USA, cordes@mlml.calstate.edu

Charles K. Curlee, Marathon Oil Co., 5555 San Felipe, Houston, TX, 77056, USA

Andre d'Entremont, Canada-Nova Scotia Offshore Petroleum Board, 6th Floor TD Centre, 1791 Barrington St., Halifax, NS, B3J 3K9, Canada, adentremont@cnsopb.ns.ca

Nancy Dahl-Tacconi, Marine Protected Areas, Environment Australia, GPO Box 787, Canberra, ACT 2601, Australia, nancy.dahl-tacconi@ea.gov.au

Derek Davis, 10 Forest Rd., Dartmouth, NS, B3A 2M3, Canada, ap775@chebucto.ns.ca

Paul Dayton, Scripps Institution of Oceanography, La Jolla, CA, 92093-0227, USA, pdayton@ucsd.edu

Ben De Mol, RCMG, University of Gent, Krijgslaan 281, S8 B-9000 Gent, Belgium, ben.demol@rug.ac.be

Jennifer Dianto, New England Aquarium, Conservation Dept., Central Wharf, Boston, MA, 02110, USA, jdianito@neag.org

Penny Doherty, Ecology Action Centre, 1568 Argyle St., Suite 31, Halifax, NS, B3J 2B3, Canada, padohert@is.dal.ca

Natasha Duchene, PO Box 550, Chester, NS, BOJ 1J0, Canada, orange.sky@never-ever.net

David R. Duggan, Fisheries and Oceans Canada, Oceans Act Coordination Office, Mail Stn. B500, PO 1006, Dartmouth, NS, B2Y 4A2, Canada, duggand@mac.dfo-mpo.gc.ca

Marie Eiland, Zoological Museum of Copenhagen University, Universitetsparken 15, DK-2100 Copenhagen, Denmark, meiland@zmuc.kd.dk

Susan Feindel, General Delivery, Chester Basin, NS, B0J 1K0, Canada, asv@mby.auracom.com

Derek Fenton, Fisheries and Oceans Canada, Oceans Act Coordination Office, PO Box 1006, Mail Stn B500, Dartmouth, NS, B2Y 4A2, Canada, fentond@mar.dfo-mpo.gc.ca

John Field, US Fish and Wildlife Service, 4401 N. Fairfex Dr., Rm 750, Arlington, VA, 22303, USA, john-field@fws.gov

Jan Helge Fosså, Institute of Marine Research, PO Box 1870, Nordnes, N-5817 Bergen, Norway, jhf@imr.no

Scott C. France, College of Charleston, 58 Coming St., Room 216D, Charleston, SC, 29424, USA, frances@cofc.edu

André Freiwald, Institut für Geologie und Paläontologie, Universität Tübingen, Herrenberger Strasse 51, 0-72070 Tübingen, Germany, andre.freiwald@uni-tuebingen.de

Susanna Fuller, Biology Dept., Dalhousie University, Halifax, NS, B3H 4J1, Canada, sdfuller@is2.dal.ca

Susan E. Gass, Ecology Action Centre, 1568 Argyle St., Suite 31, Halifax, NS, B3J 2B3, Canada, coral@is.dal.ca

Robert Y. George, University of North Carolina at Wilmington, Wilmington, NC, 28403, USA, georger@unowil.edu

Brian Giroux, Scotia Fundy Mobile Gear Fisherman's Association, 33 Chestnut St, Yarmouth, NS, B5A 2N7, Canada

Donald Gordon, Habitat Ecology Section, Bedford Institute of Oceanography, PO Box 1006, Dartmouth, NS, B2Y 4A2, Canada, gordond@mar.dfo-mpo.gc.ca

Alex Green, National Sea Products, 100 Battery Point, PO Box 910, Lunenburg, NS, BOJ 2C0, Canada

Edmund Green, UNEP World Conservation Monitoring Centre, 219 Huntingdon Rd., Cambridge, CB3 0DL, UK, ed.green@une-wcmc.org

Anthony Grehan, Martin Ryan Marine Science Institute, National University of Ireland, Galway, Ireland, anthony.grehan@nuigalway.ie

Richard Grigg, Dept. of Oceanography, University of Hawaii, 1000 Pope Rd., Honolulu, HI, 96822, USA, rgrigg@soest.hawaii.edu

Bruce C. Hatcher, Dept. of Biology, Dalhousie University, Halifax, NS, B3H 4S1, Canada, bhatcher@is.dal.ca

Christopher Hawkins, Triton Consultants, 221 Crichton Ave., Dartmouth, NS, B3A 3R8, Canada, hawkinscm@accesswave.ca

Andrew Hebda, Nova Scotia Museum of Natural History, 1747 Summer St., Halifax, NS, B3H 3A6, Canada, hebdaaj@gov.ns.ca

Phillip Heck, Geological Institute, Swiss Federal Institute of Technology, Baslerstrasse 125, CH-8048 Zurich, Switzerland, phillip@darksky.ch

Doug Hollett, Marathon Canada, 1959 Upper Water St., Suite 401, Halifax, NS, B3J 3N2, Canada, dwholett@marathonoil.com

Martin Hovland, Staoil, Forusbeen 50, N-4035 Stavanger, Norway, mhovland@statoil.com

David Humphry, Lower Clarks Harbour, Shelburne County, NS, BW0 1PO, Canada

Walter C. Jaap, Florida Marine Research Institute and Lithophyte Research, 273 Catalan Blvd., NE Saint Petersburg, Florida, 33704, USA, wkjaap@worldnet.net.att

Sharon Jeffery, Vancouver Aquarium Marine Science Centre, PO Box 3232, Vancouver, BC, V6B 3X8, Canada, jeffers@vanaqua.org

Christopher Jones, Department of Fisheries and Oceans, 176 Portland St., 5th floor, Marine House Dartmouth, NS, B2Y 1J3, Canada

Derek Jones, Canadian Ocean Habitat Protection Society, Box 13, Newellton, NS, B0W 1P0, Canada, dkpjones@atcon.com

Lisbeth Jonsson, Tjearno Marine Biological Laboratory, SE-452 96 Stromstad, Sweden, lisbeth.jonsson@tmbl.gu.se

Ellen Kenchington, Department of Fisheries and Oceans, 1 Challenger Drive, Dartmouth, NS, B2Y 4A2 Canada, kenchingtone@mar.dfo-mpo.gc.ca

Trevor Kenchington, Gadus Associates, RR#1, Musquodoboit Harbour, NS, B0J 2L0, Canada, gadus@istar.ca
J. Anthony Koslow, CSIRO Marine Research, GPO Box 1538, Hobart, Tasmania 7001, Australia, tony.koslow@marine.csiro.au
Vladimir Kostylev, Marine Environmental Geosciences, Geological Survey of Canada (Atlantic), Bedford Institute of Oceanography, 1 Challenger Dr., PO Box 1006, Dartmouth, NS, B2Y 4A2, Canada, kostol@agc.bio.ns.ca
Kenneth Krieger, National Marine Fisheries Service, Auke Bay Fisheries Laboratory, 11305 Glacier Hwy., Juneau, AK, 99801-8626, USA, Ken.krieger@noaa.gov
Marc S.S. Lavaleye, Netherlands Institute for Sea Research (NIOZ), PO Box 59, 1790 AB Den Burg, The Netherlands, lava@nioz.nl
Tomas Lundalv, Tjaernoe Marine Biological Laboratory, SE-452 96 Stroemstad, Sweden, tomas.lundalv@tmbl.gu.se
Kevin MacIsaac, Marine Environmental Sciences Division, Department of Fisheries and Oceans, Bedford Institute of Oceanography, PO Box 1006, Dartmouth, NS, B2Y 4A2, Canada, macisaack@mar.dfo-mpo.gc.ca
Paul Macnab, Fisheries and Oceans Canada, Oceans Act Coordination Office, Mail Stn. B500, PO Box 1006, Dartmouth, NS, B2Y 4A2, Canada, macnabpcmar.dfo-npo.gc.ca
Lucie Marquardt, Alfred Wegener Institut for Polar and Marine Research, Columbusstrasse, 27568 Bremerhaven, Germany, lmarquardt@awi-bremerhaven.de
Mona McCrea, Ecoserve, 17 Rathfarnham Rd., Terenure, Dublin 6W, Ireland, mona@ecoserve.ie
Denise McCullough, 3242 Agricola St., Halifax, NS, B3K 4H3, Canada, dcoast@hfx.eastlink.ca
Cathy Merriman, WWF Canada, 245 Eglinton Ave. East, Suite 410, Toronto, ON, M4P 3J1, cmerriam@wwfcanada.org
Anna Metaxas, Dept. of Oceanography, Dalhousie University, Halifax, NS, B3H 4J1, Canada, metaxas@is.dal.ca
Christopher Miller, Dept. of Biology, University of Waterloo, Waterloo, ON, N2L 3G1, ca7mille@scimail.uwaterloo.ca
Anthony D. Montgomery, The Oceanic Institute, 41-202 Kalanianaole Hwy., Waimanalo, HI, 96795, USA, amontgom@hawaii.edu
James P. Mortimor, International Centre for Island Technology, Heriot-Watt University, Old Academy, Back Road, Stromness, Orkney, KW16 3AW, Scotland, j_mortimor@hotmail.com
Ransom Myers, Dept. of Biology, Dalhousie University, Halifax, NS, B3H 4J1, Canada, myers@mscs.dal.ca
Abdulla Naseer, Dept. of Biology, Dalhousie University, Halifax, NS, B3H 4J1, Canada, anaseer@is2.dal.ca
Dianne Nickerson, 2703 Fuller Terrace, Halifax, NS, B3K 3V9, Canada
Elliott Norse, Marine Conservation Biology Institute, 15806 NE 47th Ct., Redmond, WA, 98052, USA, elliott@mcbi.org

Leif Nottestad, Institute of Marine Research, PO Box 1870, Nordnes, N-5817 Bergen, Norway, leif.nottestad@imr.no
Covadonga Orejas, Alfred Wegener Institute for Polar and Marine Research, Columbusstrasse, 27568 Bremerhaven, Germany, corejas@awi-bremer-haven.de
Christine Penney, Clearwater Fine Foods Inc., 757 Bedford Hwy., Bedford, NS, B4A 3Z7, Canada, cpenney@cffi.com
Alan Pinder, Dept. of Biology, Dalhousie University, Halifax, NS, B3H 4J1, Canada, pinder@is.dal.ca
Tom Pinfold, Gardner Pinfold Consulting, 1331 Brenton St., Halifax, NS, B3J 2K5, Canada, tpinfold@gardner.ns.ca
Ted Potter, Oceans Act Coordination Office, Fisheries and Oceans Canada, PO Box 1006, Mail Stn. B500 Dartmouth, NS, B2Y 4A2, Canada, pot-tert@mar.dfo-mpo.gc.ca
John Reed, Harbour Branch Oceanographic Institution, 5600 US 1 North, Fort Pierce, Florida, 34946, USA, jreed@hboi.edu
Virginia Rich, University of Washington Seattle, General Delivery, Corea, ME, 04624, USA, vrich@alum.calbekeley.org
Michael Risk, Dept. of Geography and Geology, McMaster University, Hamilton, ON, L8S 4M1, Canada, riskmj@mcmail.cis.mcmaster.ca
Gonzalo Macho Rivero, Centro de Investigacions Marifias, Xunte de Galicia, Apdo 13, Vilanova de Arousa, Pontevedra 36620, Spain, gma-cho@correo.unigo.es
Murray Roberts, Scottish Association for Marine Science, PO Box 3, Oban, Argyll, Scotland, PA3A 4AD, UK, m.roberts@dml.ac.uk
Jayne Roma, Ecology Action Centre, 1568 Argyle St., Suite 31, Halifax, NS, B3J 2B3, Canada, romaj@mar.dfo-mpo.gc.ca
Alan Ruffman, Station M, Halifax, NS, B3J 2L4, Canada
Robert J. Rutherford, Fisheries and Oceans Canada, Mail Stn. B500, PO Box 1006, Dartmouth, NS, B2J 4A2, Canada, rutherfordb@mar.dfo-mpo.gc.ca
Hector Reyes Bonilla, University of Miami, Rosenstiel School of Marine and Atmospheric Science, 4600 Rickenbacker Causeway, Miami, FL, 33149, USA, hreyes@rsmas.miami.edu
Heidi Schaefer, 153 Spring St., Summerside, PEI, C1N 3G2, Canada, hlschae-fer@hotmail.com
Robert Scheibling, Biology Dept., Dalhousie University, Halifax, NS, B3H 4J1, Canada, rescheib@is.dal.ca
William W. Schroeder, University of Alabama, Dauphin Island Sea Lab, 101 Bienville Blvd., Dauphin Island, AL, 36528, USA, wschroed@jag-uar1.usouthal.edu
David B. Scott, Centre for Marine Geology, Dalhousie University, Halifax, NS, B3H 3J5, Canada, dbscott@isdal.ca
Edward Seidel, Monterey Bay Aquarium, 886 Cannery Row Monterey, CA, 93940, USA, eseidel@mbayaq.org

Anne Simpson, University of Maine, Darling Marine Centre, 193 Clarks Cove Rd., Walpole, ME, 04573, USA, anne.simpson@_umit.maine.edu

Jodie Smith, Dept of Geography and Geology, McMaster University, Hamilton, ON, L8S 4M1, Canada

Margaret Spring, Democratic Senior Council, Oceans and Fisheries Subcommittee, Committee on Commerce, Science and Transportation, 566 Dirksen Senate Office Bldg., Washington, DC, 20510, USA, margaret_spring@commerce.senate.gov

Robert Stone, National Marine Fisheries Service, Alaska Fisheries Science Center, 11305 Glacier Hwy., Juneau, AK, 99801-8626, USA, bob.stone@noaa.gov

Mike Strong, Fisheries and Oceans Canada, 531 Brandy Cove Rd., St. Andrews, NB, E5B 2L9, Canada, strongm@mar.dfo-mpo.gc.ca

Sarah C. Stuart, Pew Charitable Trusts, 2005 Market St., Philadelphia, PA, 19103, USA, sstuar@pewtrusts.cpm

Tim Surette, Department of Fisheries and Oceans, 215 Maine St., Yarmouth, NS, B5A 1C6, Canada, surettet@mar.dfo-mpo.gc.ca

Tomas Tomascik, Parks Canada –NCSC, 300-300 West Georgia St., Vancouver, BC, V6B 6B4, Canada, tomas_tomascik@pch.gc.ca

Leonardo Tunesi, ICRAM - Central Institute for Applied Marine Research, Via di Casalotti, 300 Roma I-00166, Italy, letunesi@tin.it

Les Watling, School of Marine Science, Darling Marine Center, University of Maine, Walpole, ME, 04573 USA, watling@maine.edu

Martin Willison, School for Resource and Environmental Studies, Dalhousie University, 1312 Robie St., Halifax, NS, B3H 3J5, Canada, willison@is.dal.ca

John Wilson, Dept. of Geology, Royal Holloway University of London, Egham, Surrey TW20 0EX, UK, j.wilson@gl.rhbnc.ac.uk

Matthew T. Wilson, Alaska Fisheries Science Center, NMFS, NOAA, 7600 Sand Point Way, NE Seattle, WA, 98115, USA, matt.wilson@noaa.gov

David Witherell, North Pacific Fishery Management Council, 605 West 4th Ave., Suite 306, Anchorage, AK, 99501-2252, USA, david.witherell@noaa.gov

Jim Wolford, Federation of Nova Scotia Naturalists, Site #1, Comp. #61, RR#3, Wolfville, NS, B0P 1X0, Canada, jww.triv@ns.sympatico.ca

Chris Zeman, Americans Oceans Campaign, 579 Hamilton Pl., Riverdale, NJ, 07675, USA, c_zeaman@e-mail.msn.com